I0500151

Eternal Spiritual Father

by Don Alexander

July 2017

Dedication:

People of every kindred, nation and tongue, who reject the **unscientific premise** that "random chance," **functioning in the total absence of any of the primordial elements,** can manipulate a state of "nothingness" into the sum total of all energy, matter and motion, **will find herein** supporting scientific facts proving that such a hypothesis is **absolutely impossible.**

Table Of Contents

Biblical content is paraphrased from the 1611 Authorized King James Version of the Holy Bible which is now public domain. Pronouns referring to God, Jesus Christ and the Holy Spirit have been capitalized by the author.

Introduction

In the year 1717 A.D. (300 years ago), the fastest mode of human transportation was riding a horse or sitting in some type of wagon, carriage, chariot or stagecoach drawn by a team of horses. The most advanced military weapons were cannons, crude single-shot muskets, bows and arrows, knives, swords, sabers, catapults, lances, spears, tomahawks, and other slaughter weapons used in hand-to-hand combat.

In less than five human lifetimes, man-kind has walked on the moon; engineered aircraft that can fly into the stratosphere at speeds exceeding 2,000 miles per hour; built and flown space vehicles capable of speeds in excess of 35,000 miles per hour; designed and built computers that can process thousands of bits of information in a fraction of one second; developed drugs that kill microbes, bacteria and viruses; designed and built thermonuclear weapons of war that are capable of wiping out the entire human race; and numerous other lofty scientific achievements that boggle the human mind.

Nevertheless, human scientists have been totally unable to answer the most basic questions concerning the origins of the primordial elements and the life forces enabling living organisms to move, feed, and reproduce. Neither can scientists explain nuclear fission or nuclear fusion to monkeys.

The life forces of plants, microbes, bacteria, animals, and humans are not composed of matter, energy or motion whereas the known universe is indeed composed entirely of the primordial elements.

It therefore follows as a simple scientific fact that the life forces of all living organisms originated outside of the universe known to humans. Moreover, hydrogen is the most plentiful primordial element making up the universe. It is also an undisputed scientific fact that no scientist (past or present) has a clue as to the origin of hydrogen.

Do spiritual dimensions exist of which scientists are totally ignorant? The answer is addressed in this literary effort.

Don Alexander July, 2017

Chapter one
In the beginning

Since it is indeed a scientific fact that the known universe is composed entirely of energy, matter and motion, it must also be undeniably true that the universe had a beginning; and had (by definition) to appear from the dark void of "nothingness." What is the scientific description of the state of "nothingness?" Answer: "The total absence of any form of energy, matter or motion."

Therefore, in the pure state of nothingness there is zero energy, matter or motion; and zero hydrogen or any of the other primordial elements. The total absence of hydrogen

within the state of nothingness would most
certainly preclude the **unproven hypothesis**
that the primordial elements other than
hydrogen are simply the byproducts of the
fusion of hydrogen into helium pursuant to
enormous heat and gravitational forces. In the
presumed state of nothingness, hydrogen
would not exist and neither would any other
form of energy, matter or motion exist to exert
either heat or pressure upon such non-existing
hydrogen.

Nonetheless, scientists promoting
evolutionary hypotheses have persuaded our
public schools to teach our children that the
universe actually originated as the result of the
"big bang." What is the big bang **hypothesis?**
Answer: "The hard core of a burned out star

(which cannot exist in a state of nothing-ness) contains such a condensed gravitational attraction that a "black hole" is created which sucks in all energy, matter and motion that crosses its "event horizon." The "sucked in" energy, matter and motion further condenses due to this ever-increasing gravitational force **(that cannot exist in a state of nothingness)** until the "sucked in" energy, matter and motion converts into a "singularity" **(an imaginary form of "pure energy")** which just keeps getting hotter and more dense (due to gravity) until this collection of pure energy **(singularity)** explodes **(the big bang)** into billions of trillions of cubic miles of primordial elements which collide with each other thereby transforming and gravitating the exploded elements into billions of stars, asteroids, comets, planets, galaxies, and miscellaneous interstellar masses."

The hypothesis of evolution further assumes that the burned out star **probably** remained from a series of former universes or "parallel universes" that collapsed due to unknown forces and that the life forces of microbes, plants, animals and humans were **obviously** "seeded" into Planet Earth by more advanced life forms inhabiting planets circling stars **"far away and long ago."**

Evolutionists are unable to come up with any cosmic facts that shed any scientific light upon the source of the primordial elements or the life forces connected with prior collapsed universes, parallel universes, or seeding of life on Earth by aliens. Neither can evolutionists explain how some "pure energy" the size of a walnut, or perhaps a single atom, exploded into billions of trillions of cubic miles of hydrogen. The specific gravity and maximum expansion of hydrogen gas is not a cosmic secret. The exact scientific impossibilities that moot the big bang creating our known universe also moots prior, collapsed and parallel universes.

The human spirit recognizes that the known universe **does exist** because the human brain experiences the essence of the universe through the physical senses of sight, hearing, smelling, touching, and tasting. The five physical senses of humans discern energy, matter and motion pursuant to the molecular

interaction between atoms making up the
primordial elements, synthesized elements, and
transmuted elements resulting from either
nuclear fission or nuclear fusion. All energy,
matter and motion existing within the known
universe is totally bound up within the
interaction between atoms, molecules and
compounds formed by the repelling and
attracting forces inherent within primordial,
synthesized and transmuted elements.

The individual spirit living within each
physical human body is not composed of
hydrogen, or helium, or nitrogen, or carbon, or
calcium, or any combination of the primordial
element nor any combination of the non-
primordial elements whereas all energy, matter
and motion (including human bodies) within

the known universe are indeed acknowledged
by every scientist to be composed of a limited
number of the inorganic elements listed on the
Periodic Table of the Elements and referred to
as iron, beryllium, sodium, magnesium,
lithium, potassium, calcium, scandium,
titanium, vanadium, chromium, manganese,
cobalt, nickel, copper, zinc, boron, carbon,
nitrogen, oxygen, fluorine, neon, aluminum,
silicon, phosphorus, sulfur, chlorine, argon,
gallium, germanium, arsenic, selenium,
bromine, krypton, rubidium, strontium,
yttrium, zirconium, niobium, molybdenum,
technetium, ruthenium, rhodium, palladium,
silver, cadmium, indium, tin, antimony,
tellurium, iodine, xenon, caesium, barium,
hafnium, tantalum, tungsten, rhenium,

osmium, iridium, platinum, gold, mercury,

thallium, lead, bismuth, polonium, astatine,

radon, francium, radium, rutherfordum,

dubnium, seaborgium, bothrium, hassium,

meitnreium, ununhexium, damstadium,

roentgerium, copemicium, ununtrium,

ununquadum, ununpentium, ununsepium,

ununoctium, lanthanum, actinium, cerium,

thorium, protactinium, praseodymium,

neodymium, uranium, promethium,

neptunium, samarium, plutonium, europium,

americium, gadolinium, curium, terbium,

berkelium, dysprosium, californium, holmium,

einsteinum, erbium, fermium, thulium,

mendelevium, ytterbium, nobelium, lutetium,

and lawrencium.

The **entire universe** is composed of

hydrogen (90.71%), helium (8.59%), carbon (0.02%), nitrogen (0.04%), oxygen (0.06%), plus trace elements (all the other elements totaling 0.58%). Earth's crust, surface, oceans and atmosphere consist entirely of oxygen (49.2%), silicon (25.7%), aluminum (7.50%), iron (4.71%), calcium (3.39%), sodium (2.63%), potassium (2.40%), magnesium (1.93%), hydrogen (0.87%), titanium (0.58%), chlorine (0.19%), phosphorous (0.11%), manganese (0.09%), carbon (0.08%), sulfur (0.06%), barium (0.04%), nitrogen (.04%), fluorine (0.03%), and trace elements (totaling 0.49%).

The human body per 70 kilograms of body weight is composed of oxygen (45.5 kg), carbon (12.6 kg), hydrogen (7.0 kg), nitrogen

(2.1 kg), calcium (1.0 kg), phosphorus (0.70 kg), magnesium (0.35 kg), potassium (0.24 kg), sulfur (0.18 kg), sodium (0.10 kg), chlorine (0.10 kg), iron (0.003 kg), zinc (0.002 kg), plus trace elements present in less than one milligram quantity for each element (arsenic, chromium, cobalt, copper, fluorine, iodine, manganese, molybdenum, nickel, selenium, silicon, and vanadium).

Consequently, the inescapable, firmly established, indisputable fact, known and 100% accepted by every scientist in every field of 21st century science, is that the human life force (the human spirit) is most definitely not composed of either the primordial or the non-primordial elements. Therefore, the human life force originated within and returns to a spiritual dimension separate and distinct from matter, energy and motion perceived by humans through the mortal senses of vision, hearing, touching, tasting and smelling within the three dimensional human habitat defined as space, distance and time. The human physical anatomy including the human brain cannot comprehend that which originates and remains within a spiritual dimension. Likewise, an earthworm cannot perceive nor communicate with the hierarchy of life above the surface of the earth.

Although exceedingly more complex in physical structure than a bacterium, an earthworm spends its entire existence in moist soil feeding on decaying organic matter. Thus, there is zero earthworm perception of lunar cycles, milky way orbit, earth orbiting satellites, war and peace among humans, rockets, missiles, fighter jets, thermonuclear bombs, etc.

Such lack of earthworm awareness of above ground life does not preclude the existence of earth's life forms ranging from bacteria to humans. Neither does the lack of human perception of spiritual dimensions and spiritual beings such as angels preclude the existence of living beings within the hierarchy of life forms as far above humans as humans are removed from common earthworms.

Chapter two
Energy, Matter and Motion

The total amount of energy, matter and
motion within the known universe does not
change. There has never been an exception to
the Newtonian Laws of Physics: (1) the
conservation of energy; (2) the conservation of
matter; and (3) energy or matter cannot be
either created or destroyed by natural
processes. Neither has there ever been an
exception to the Law of Thermodynamics
relating to the temperature equilibrium of the
universe: When the temperature of the
universe reaches absolute zero, all molecular
interaction will totally cease.

In other words, the mount of energy **added to an open system is equal to the amount of heat added less the amount of heat lost to the universe.** The amount of heat escaping into the universe is an irreversible transfer of heat and is known as entropy. The Law of Entropy states that when heat escapes into the universe the level of disorder in the universe increases because the energy transfer is **irreversible** such that disorder within the universe is also irreversible which dictates that disorder **never** evolves into a more ordered state as a naturally occurring process. More-over, the Law of Cause and Effect plus the Newtonian Laws of Physics, plus the Laws of Thermodynamics, plus the Law of Entropy 100% moot the "Big Bang" hypothesis and the concept of the spontaneous generation of original life forms; and the hypothesis that bacteria evolved into humans over billions of years.

This scientific analysis, of course, begs the question: Where, when and how did the original energy, matter and motion which makes up the known universe come into being? It is for certain that energy, matter and motion did not suddenly emerge as a big bang from a state of nothingness because energy and matter cannot be either created or destroyed as naturally occurring processes. Energy, matter and motion can only be transformed, by the application of heat, from plasma to gas to liquid to solid molecular structures with some heat escaping into the universe thereby adding

to the irreversible state of disorder in accordance with the Law of Entropy.

"Energy" is actually "matter in motion"
and "motion" is "evidence that energy is inter-
acting with matter." When energy interacts
with matter, a force results. There exists four
primary forces within the known universe:
gravity, the strong nuclear force, the weak
nuclear force, and the electromagnetic force.

Individual atoms of individual elements
except hydrogen are composed of one or more
electrons plus one or more protons. Most of
the elements contain one or more neutrons
within the nucleus of an individual atom of the
element under consideration. Protons repel
other protons. Protons attract electrons, and
neutrons neither attract nor repel other

neutrons, protons or electrons.

The strong nuclear force keeps the repelling protons from tearing the atom apart. An atom with a weak nuclear force allows the **loss of one or more protons.** An atom of an individual element that loses one or more **protons** transmutes into an atom of a **different** individual element. During transmutation, the atom under consideration releases an electro-magnetic force field. The most common electromagnetic force field is **sunlight** which results from the transmutation of hydrogen into helium pursuant to nuclear fusion within the masses of matter known as stars.

Gravity is the attracting force field that is inversely proportional to the distance between the masses under consideration. Two or more atoms of an individual element form a molecule of the element and the molecule is held together by the bonding attraction between electrons and protons. Atoms and/or molecules of **different elements** can also be bonded together by the attracting force between electrons and protons; and the resulting molecule pursuant to such bonding is referred to as a compound. For example, when chlorine is bonded with sodium, the compound known as common table salt results. When hydrogen gas is bonded with oxygen gas, the resulting compound is known as liquid water.

"Quantum Mechanics" is the scientific field of study that explores molecular structure within individual elements at the microscopic level. For example, if counting four electrons

per second around the clock, it would take
more than nineteen million years to count the
number of electrons in a line one inch wide.
How many electrons would be contained in 14
billion times 6 trillion cubic miles of com-
pounded molecular matter? Answer: The
number ten to the eightieth power.

What are the odds against a "primordial
soup" of individual primordial elements
creating by random chance the DNA code for
the simplest living organism known to modern
science? Answer: The simplest microbe
known to modern scientists is "Mycoplasma
genitalium" which contains 525 genes in its
genome (complete set of DNA base pairs
including genes). According to bio-engineers
at Stanford University who conducted a

simulation of a single reproductive cycle of this simplest free-living bacterium, the computerized model for one single division of the bacterium required 28 integrated sub-systems using 128 computers running for 10 hours and generating one half gigabyte of data.

Sir Fred Hoyle, a famous evolutionist with expertise in mathematical probabilities, calculated that the odds against the protein building blocks necessary for DNA coding **(making possible one reproductive cycle of a single-celled organism through mitosis)** compounding by happenstance are one chance in ten to the billionth power. Hoyle also calculated the odds of basic protein "building blocks" (necessary for human DNA) forming by happenstance to be one chance in ten to the forty thousandth power. Never, in the entire history of human research and scientific documentation has any event occurred where the odds against such occurrence were one chance in ten to the **fiftieth power.** Compare this tidbit of scientific certainty to ten to the eightieth power which approximates the total number of electrons existing within the known universe.

Earth's orbital velocity is roughly 67,000 miles per hour and Earth's axial velocity is around 1,040 miles per hour. The Solar System containing Earth orbits the Milky Way Galaxy at approximately 514,000 miles per hour. The Milky Way is receding from neighboring galaxies at an estimated velocity of six hundred and seventy million, three hundred and twenty thousand miles per hour (65 % of the speed of light). Traveling at the velocity of light, it would take 100,000 Earth years to cross the maximum diameter of the Milky Way which is a modest sized galaxy among the billions of galaxies existing within the known universe and together containing hundreds of billions of stars and billions of planets. Earth's sun represents slightly more than 99.0 % of the total matter composing the solar system containing Earth, which is the only planet supporting a variety of living organisms within the entire known universe.

Intelligent design, engineering, and exquisite creativity are extremely obvious to human observers of the universe on the macro scale. The micro universe is even more precise and finely balanced within space, distance and time. Within an individual atom of an individual element, each electron orbits the nucleus of the atom completing billions of revolutions per millionth of a clock second. The astounding electron speed around the atom nucleus makes the electron appear to vibrate rather than orbit.

Each proton has a mass equal to the mass of 1,836 electrons. The mass of an electron is 0.0000000000000000000000009 of one gram which is very close to zero mass. It takes 1,839 electrons to equal one neutron's mass. The protons and neutrons are confined within the tiny nucleus of the atom and are in constant motion. The higher the velocity, the higher the temperature. Every element has a freezing point, melting point and boiling temperature. At extremely high temperatures some elements convert to plasma as compared to a gaseous state. Thus, all elements exist, based upon temperature and pressure, as plasma, gas, liquid or solid.

Atoms can gain, lose or share electrons during chemical reactions with atoms of another element. An atom that loses one or more electrons becomes a positive ion whereas an atom that gains one or more electrons becomes a negative ion.

A negative or positive ion results when the negative charge on the atom's total electrons does not match the positive charge on the atom's total protons.

If an atom easily gives up electrons, its

valence is positive, and atoms that tend to gain electrons have a negative valence.

Sodium tends to lose its one electron and thus has a valence of (+1). Chlorine tends to accept one electron from another atom and therefore has a valence of (-1). Negative ions can chemically bond with positive ions.

Thus, a molecule of ordinary table salt consists of one atom of sodium linked to one atom of chlorine. This type of chemical interaction between the atoms of the known elements is how all the matter in the universe, is structured.

The nucleus makes up nearly all the mass of an atom. Protons and neutrons which make up the nucleus are roughly 100,000 times smaller than the atom. Electrons are not known to be composed of smaller particles of matter whereas protons and neutrons are composed of smaller particles called quarks.

Each proton and each neutron is made up of three quarks. Quarks can be manipulated by researchers within a science laboratory to form other particles of matter besides protons and neutrons but such particles are highly unstable and break down within a tiny fraction of a second. Therefore, these unstable particles are not found outside the laboratory.

Each electron has inherent energy in proportion to its orbiting velocity. The **strong nuclear force** binding protons within the atom's nucleus also, **like electrons,** appears to vibrate rather than orbit and is believed to be the actual source of gravitational attraction between masses.

The positively charged protons within the nucleus exert a force on orbiting negatively charged electrons that keeps them within the atom when the atom is not involved in a chemical reaction, nuclear fission, or nuclear fusion. The inherent energy within an electron generates resistance to the attracting force of the nucleus. The more energy the electron has, the farther from the nucleus it will be.

Consequently, electrons are arranged in energy shells at varying distances from the nucleus as determined by the level of their inherent energy. Electrons with the least energy are located in the inner shells and those with higher energy levels are in the outer shells.

Each electron energy shell is identified
by a number or letter. The shell closest to the
nucleus is shell #1 or shell K. The other shells,
in order of increasing distance from the
nucleus, are numbered 2 through 7 or labeled L
through Q. Each shell can hold a limited
number of electrons. Shell 1 can hold no more
than 2 electrons. Shell 2 can hold 8 electrons,
shell 3 can hold 18, shell 4 can hold 32, shell 5
can hold 50, shell 6 can hold 72, and shell 7
can hold 98. However, the outer shells are
never completely filled. The number of filled
shells is determined by the number of electrons
contained within the atom. An atom that has
lost all its electrons will become a positively
charged free nucleus.

There can also be free electrons

(negative charge), and free neutrons (neutral charge) as the result of radioactive decay, nuclear fission and nuclear fusion. In the atoms of radioactive elements the nucleus will change as the atom gives off radioactive particles.

The change in the nucleus may be the simple rearrangement of its protons and neutrons or the actual loss of one or more. If only the arrangement of the nucleus changes, gamma rays are emitted from the atom. If the number of protons changes, alpha or beta radiation is given off. When an atom loses one or more protons, it transmutes to an atom of a different element.

If one or more neutrons escape from the nucleus, the atom becomes an isotope of the radiating element. All elements heavier than bismuth are radioactive as well as the isotopes of some of the lighter elements. Isotopes of nearly all the elements can be created by bombarding their atoms with subatomic particles.

The atomic number denotes how many protons an atom of an element contains, and the mass number identifies the sum of the protons and neutrons within the nucleus. Atomic weight is the weight of an atom expressed in "atomic mass units" (amu). One amu or "dalton" equals 1/12 the weight of an atom of carbon 12. There are 602 billion trillion amu in one gram.

All atoms of the same element have the same number of protons. Since every hydrogen atom contains one proton, the atomic number of hydrogen is (1). The atomic numbers range successively up to 94 for plutonium because this element has 94 protons in each atom. Elements with more than 94 protons in each of its atoms can be created by scientists in the laboratory.

There exists more than one isotope for most of the elements. For example, hydrogen has three. The most common has no neutron in the nucleus of each atom. In the other two isotopes, the nucleus contains one to two neutrons. The mass number is used to distinguish the three isotopes; hydrogen 1, hydrogen 2 and hydrogen 3. These isotopes are also called protium, deuterium and tritium respectively.

Most of the lighter elements contain about the same number of protons and neutrons in the nucleus of their atoms. The heavier elements have more neutrons than protons. The heaviest elements have about three neutrons for every two protons. U-238, for example, has 92 protons and 146 neutrons.

Atoms of different elements which have the same mass number but different atomic numbers are called isobars. The isobars argon and calcium have a mass number of 40 but argon's atomic number is 18 (18 protons) and calcium's atomic number is 20 (20 protons).

The way an atom of an element behaves during a chemical reaction is largely determined by the number of electrons in its outermost electron shell. When atoms combine and form molecules, electron(s) in the energy shell of each atom **which is/are the greatest distance from the nucleus** is/are either transferred from one atom to another or shared between atoms.

The number of electrons involved in the chemical reaction is referred to as "valence." The atoms of some elements can have more than one valence depending on the number and kind of atoms they can combine with.

Electrons are restricted to a limited set of motions, each of which has a specific energy value. These motions are referred to as quantum states or energy levels. When an electron is in a given quantum state, it does not give off or absorb energy. An atom can lose or gain energy only when one or more electrons change from one quantum state to another.

Electrons seek the lowest state of energy but only one electron at a time can occupy each quantum state. When the lower states are filled, other electrons are forced to occupy higher states. When all electrons are in the lowest available state, the atom is in "ground state" which is the normal condition for atoms at ordinary temperatures.

When matter is heated a few hundred degrees, sufficient energy is then available to raise one or more electrons to a higher energy level. The atom is then transformed into an "excited state" which lasts for a fraction of a second. An excited electron quickly drops to a lower state and continues dropping until the atom returns to its ground state.

During each succeeding drop, the

electron gives off a tiny packet of radiant energy called a photon. The energy of the photon equals the difference in energy between the two energy levels the electron passed through. These photons are detected as visible light and other forms of electromagnetic radiation.

One neutron and one proton can occupy each quantum state in the nucleus of an atom. A light nucleus has about the same number of protons and neutrons but a proton and neutron in the same state do not have the same amount of energy because each proton is electrically repelled by all other protons in the nucleus thereby increasing the energy of each proton.

In a heavy nucleus, the difference in energy levels between protons and neutrons is significant and more low energy states are available for neutrons than for protons. This helps explain why a heavy nucleus contains more neutrons than protons.

Most of the 94 elements found in and on Earth (as contrasted to the elements created in scientific laboratories and nuclear reactors) are in compound form. They are combined with other elements forming soil, rocks, gas, liquids, minerals, crystals, etc. Oxygen and silicon are the most plentiful elements in Earth's crust and make up 3/4 of the crust's weight. A few elements are found in pure form in small amounts such as gold, copper, carbon and sulfur.

It is easy to confuse energy, force and power. Energy is the word used to describe the ability to make things happen, like raising the temperature of liquids, gases and solids. Energy propels, directs and accelerates all types of matter; produces light; binds sub-atomic particles within the nucleus of all atoms; and numerous other activities classified as "work." The amount of activities which can be accomplished depends upon the strength of the energy force used and the distance through which it moves. Power measures the rate at which the work is performed.

All matter is held together by the energy which prevents the nucleus of every atom from self destructing. Therefore, energy existed within the universe before any of the elements came into being. An atom of any of the elements is mostly empty space invaded by energy emitting from the tiny nucleus and the orbiting electrons.

The incredible amount of energy present within the strong nuclear force which offsets the repelling force of the positively charged protons is measured by Einstein's formula: energy equals mass times the speed of light squared ($E = M$ times C squared).

The destructive power of nuclear weapons results from releasing the strong nuclear force from within the atoms of certain radioactive elements.

Since energy preceded the formation of matter and matter is composed of atoms, atoms were perfectly designed so that energy could hold them together. In other words, atoms had to come into existence within the universe in the form of their existing irreducible level of complexity.

Although closely intertwined, energy and matter are not the same. It is obvious from the very structure of matter that it is mostly energy in motion. Yet, the force which holds matter together is not matter; it is energy. Energy and matter were applied to a specific molecular design and living beings were introduced into a physical environment that had already been carefully crafted to nurture and sustain them.

The questions that physicists, chemists, astronomers, and biologists wrestle with are extremely basic but not simple. How did the universe come into existence? Was the birth of the universe accidental or designed? When did

the universe appear? Is the universe progress-
ing from a beginning to an end? Which
elements make up the bulk of the matter
contained in the universe? How did this
combination of elements originate? How were
living organisms introduced into the universe?
What forces account for the delicate balance
between energy, matter, and momentum?
Where did viruses and bacteria come from?
Where and how did the hierarchy of bacteria,
insects, plants, animals and humans originate?
How do inorganic compounds acquire life?
What exactly is life? Is there life after physical
death for humans? For other living organisms?
How and when does the human fetus acquire
life? Does a human life force contain a form of
energy? If so, what kind of energy?

What in is the substance of energy? Scientifically speaking, all energy is the gravitational, electromagnetic, and nuclear force fields created by the constant molecular interaction bound up within the atoms of the elements thereby creating potential, kinetic and heat energy.

A human body is composed of the same elements as dirt, water, and other inorganic compounds that do not contain the life force. That is an absolutely established fact and not disputed even by Darwin's disciples.

But, what then does the life force consist of? With that puzzling issue put aside for the moment, what about the universe itself? Is it not also composed of the elements just like a physical human body but in different

compounds? All physicists agree on this point.
Then, it is certain that the elements had to be
in existence before the universe could emerge
in either a primordial form or the form we
behold today. There is no other explanation
being voiced in opposition to the premise that
the elements (most certainly hydrogen and
helium) had to exist prior to what they
indisputably combined to create through
nuclear fusion and nuclear fission.

Then does it not also follow that the
most primitive form in which the universe ever
existed, whether as parallel universes or
collapsed universes (or universes that
languished for unknown eons of time within
outer darkness) must have also been formed
by the primordial elements (before light began

emitting from billions of galaxies filled with stars including our sun)? Are not the sun and other stars giant circular clouds of hydrogen and lesser amounts of other elements spinning in space and ignited by the heat generated by total inherent mass and internal molecular energy emitting from the fusion of hydrogen into helium?

So then, common logic dictates that some of the elements existed prior to the heavens and Earth and interstellar masses of whatever size, configuration and elemental composition.

Hierarchy of Life Forms
Chapter three

Where, when and how did the known
universe consisting of energy, matter and
motion first become discernible to the human
senses of vision, hearing, touching, tasting and
smelling? The answers to this three-part
question cannot be communicated to human
beings for the same reasons that jetliners,
spaceships, hydrogen bombs, and humans
cannot be communicated to earthworms. The
intellectual capacity of humans, confined to a
three dimensional universe limited to space,
distance and time, simply cannot embrace
spiritual dimensions inhabited by immortal

living beings whose intellectual and physical powers are not limited by the molecular structure of energy, matter and motion.

To postulate that "nothingness" evolved into the known universe pursuant to the total absence of energy, matter and motion producing a "big bang" is patently moronic and intellectually insulting. Both the macro and the micro universe reflect astounding intellect, engineering and indescribable creativity pointing to a supreme immortal being inhabiting multiple spiritual dimensions and exercising creative capabilities completely incomprehensible to mortal humans inhabiting a single planet within a three dimensional physical universe composed entirely of energy, matter and motion.

Scientists describe matter as the substance or substances of which any physical object consists or is composed and which exists as a plasma, gas, liquid or solid; or, the substances of which earth is composed having mass and occupying space and is distinguished from an incorporeal substance such as spirit or mind; or from qualities, actions, thoughts and emotions. (Google, 2017)

Matter, other than single-proton hydrogen, cannot exist in the absence of the energy which keeps protons from tearing the atomic nucleus apart. Neither can energy exist in the absence of the atomic particles of matter comprising protons, electrons and neutrons. Motion cannot exist in the absence of energy originating within one or more of the four primary forces known as gravity, the strong nuclear force, the weak nuclear force, and the electromagnetic force. Consequently, when the supreme, immortal, creative being (referred to as "God" by humans) conceived, engineered, and commanded the known universe to materialize; matter, energy and motion appeared in simultaneous unity.

At the very bottom of the hierarchy of life forms are single-celled living organisms classified as "bacteria." The bacterium E. coli's genome consists of a single DNA molecule containing 4,639,221 base pairs arranged in a precise chronological order and encoding 4,288 proteins. In the absence of its DNA molecule, E. coli, nor any other bacterium, cannot replicate which means without any question whatsoever that DNA coding for bacteria was genetically designed and engineered prior to the first replication of any single-celled bacterium upon Planet Earth.

This also dictates that it is genetically impossible for any living organism within the hierarchy of life forms to evolve from a lower life form pursuant to spontaneous generation combined with random chance genetic mutations occurring over eons of time producing reptiles and mammals from bacteria ancestors. The hypotheses of "natural selection" coupled with "survival of the fittest" enabling alligators to evolve into birds and catfish into whales stems from the genetic ignorance of scientists pontificating in the 19th

century about molecular and biological facts of
which they were totally ignorant.

Random chance progression of life
forms from vegetation to bacteria to marine
life to amphibians to reptiles to birds to
mammals to humans through billions of
minute random chance genetic mutations over
millions of years is **not supported by a
single proven fact** within the sum total of
accumulated human knowledge. Natural
selection and survival of the fittest through
mutational happenstance is sheer nonsense
because accidental genetic mutations
ultimately responsible for extremely complex
and beneficial biological abilities, not being
coordinated and operational until millions of
years into the future, would have to survive

certain natural extinction of non-beneficial mutations spanning millions of generations (the very core of Darwinian logic).

For example, all the components necessary for vision would have to derive from **virtually simultaneous** random chance genetic mutations to avoid extinction through the process of "natural selection" (making possible the "survival of the fittest") in the absence of DNA coding which would also have to be birthed by spontaneous generation against odds of one chance in ten to the billionth power occurring successfully more than four million times in succession exemplified by the single-celled gram negative bacterium E. coli which contains 4,639,221 base pairs of nucleotides in a very precise chronological chain of synthesized proteins.

Random chance genetic mutations do in fact occur but invariably are minor deviations in the actual execution of DNA coding. Such mutations are rarely beneficial to the replicating organism with the possible exception of molecular interaction between antibodies and a pathogenic virus or bacterium. A significant genetic mutation during the reproductive cycle would result in crippling deformity or death of the offspring.

To postulate that minor random chance genetic mutations occurring over millions of generations would produce a superior living organism molded by "natural selection" and "survival of the fittest" in the total absence of design, order or purpose is rank imagination mingled with intellectual dishonesty. How many quadrillion random chance genetic mutations distributed over millions or billions of years would be required to develop the five senses we know as vision, hearing, touching, smelling and tasting?

All the physical components of vision or hearing or touching or tasting or smelling must be possessed by the organism or living creature at the same time in order for the sense to physically function. "Natural selection" and "survival of the fittest" would have extincted the individual mutants long before the random chance mutations could be synchronized by such happenstance into any one of the five senses not to mention highly complex bodily functions such as metabolism, waste excretion, cellular mitosis, hormonal interdependence, the electron transport system, etc.

How many trillions of fossils should be buried in the earth's crust documenting the progression of bacteria into humans? Why have only a measly few fossils been hailed as "missing links" and later discovered to be "planted" by die-hard evolutionists or simply misidentified?

Honkers of Darwinian evolutionary hypotheses claim that "mountains of scientific evidence" prove that humans first appeared on Planet Earth approximately 200,000 years ago. Currently, the global doubling time cycle for humans is proclaimed by Darwin worshiping scientists to be approximately sixty-three years. It is probably true that the doubling time cycle for humans was much longer 200,000 years ago. To show the absolute intellectual dishonesty of Darwinian honkers, let us assume that **between 200,000 years ago and today** the human doubling time cycle averaged **one thousand years.** That would calculate to 200 doubling time cycles between 200,000 years ago and the present time. The number two raised to the 200th power equals one trillion, 160 billion, 693 million, 804 thousand, 4 hundred, times one trillion, times one trillion, times one trillion, times one trillion. On the other hand, if the global human doubling time cycle averaged out at 183 years over the past 6,000 years, then the population of Planet Earth today would be 7.33 billion. The global human population of Earth is currently approximately 7.5 billion.

Regardless of **when** each individual plant, bacterium, fish, amphibian, reptile, bird, mammal, or other replicating organism, **whether or not currently extinct,** appeared on Planet Earth within the hierarchy of living creatures; it is an inescapable scientific fact that molecular design, biological engineering and specific DNA coding preceded the **first appearance** on Earth of every living organism described within the hierarchy of living creatures ranging from bacteria to humans.

Within the incredibly small number of so-called "missing links" misidentified or simply planted and then dug up by die-hard evolutionists, recent scientific examination documents rampant intellectual dishonesty. For example, consider the following: The fossil named Archaeopteryx was once hailed as a transition between reptiles and birds because it supposedly had teeth and claws on its wings (it was later discovered to be a hoax).

"Piltdown man" (Eoanthropus Dawson) was disturbed from eternal rest in 1912. He was presented by the evolutionists who conveniently discovered him as a missing link between man and ape. More than 500 scientific essays were penned over four decades extolling the striking similarities between ape and man exemplified in the form of Piltdown man. He was a curious fossil. He consisted of two human skulls, an orangutan jaw, an elephant molar, a hippopotamus tooth, and a canine tooth from a chimpanzee. The skulls had been treated with acid, and the other "remains" were stained with an iron sulfate solution. The canine tooth was painted and the molars were filed down.

The orangutan jaw was modified to hide the fact that the jaw did not belong to a human skull. This concoction was strewn around a quarry in Piltdown, England for later discovery as the long awaited missing link. The individuals linked to the "discovery" were famous evolutionists with impeccable credentials. Hence, no scientist bothered to closely examine Piltdown man for forty one years. When the shameless hoax was finally uncovered in 1953, Sir Kenneth Oakley found the human skulls to belong to Ona Indians and the other remains were properly identified as to origin.

"Ramapithecus" was widely acclaimed by evolutionary intellectuals as a "direct ancestor of humans." This fossil is now easily identified as an extinct cousin of orangutans.

"Nebraska man" turned out to be a fraud based on a single tooth from a rare pig.

"Java man" consisted of pieces of bone, a skull cap and three teeth scattered over a wide area and dug up over twelve months. Today, we know the bones came from a human burial site; the femur is considered human; and the skull cap is believed to be from an ape.

"Neanderthal man" was promoted as a stooped ape-man. The fossil was eventually discovered to have been formed from a diseased primate.

"Australopithecus afarensis" or "Lucy" promised hope for a "missing link" find. Non-biased examination of Lucy's inner ear, skull and bones determined Lucy to be a pygmy chimpanzee with an upright stance.

"Homo erectus" fossils have been found throughout Earth. The fossils are human in origin and reflect individuals of small stature with proportionally smaller head and brain cavity, but within the range of people today. Middle ear studies show Homo erectus to be human. The fossils have been found in close proximity to other humans.

"Australopithecus africanus" and "Peking man" were hailed for years as true missing links but are now considered simply

Homo erectus.

"Homo habilis" is now generally considered to be comprised of fragments from other fossils such as Homo erectus and Australopithecus.

"Toumai" is presented by promoters to be the "earliest member of the human family" found thus far. A number of scientists examined the fossil and identified it as coming from an ordinary ape (October 2002).

No open minded, unbiased person with sufficient scientific education to offer an intelligent opinion concerning the origin of the universe and life on Earth supports the theory of evolution as the following quotes make crystal clear:

"The pathetic thing is that we have

scientists who are trying to prove evolution which no scientist can ever prove." (Nobel prize winning physicist Robert A. Millikan)

"The theory of evolution is one of the strangest phenomena of humanity; it is entirely destitute of proof." (World famous geologist from Canada, Sir William Dawson)

"The Darwinian theory of descent has not a single fact to confirm it in the realm of nature. It is not the result of scientific research, but purely the product of imagination." (Professor Fleischmann, zoologist, University of Erlangen)

"There is not the slightest evidence that any of the major [animal] groups arose from any other." (Dr. Austin H. Clark, world famous American biologist)

"Darwin's theory of natural selection has never had any proof..." (Dr. Richard Goldschmidt, Professor of zoology, University of California)

"The Darwinian approach has consistently been to find some supporting fossil evidence, claim it as proof for evolution," and then ignore all the difficulties. It is, in fact, a common fantasy..." (Roger Lewin, science journalist)

Hypotheses based upon random chance genetic mutations are summed up by the renowned British evolutionist Sir Arthur Keith: "Evolution is unproved and unprovable. We believe it because the only alternative is special creation, and that is unthinkable."

Leaving behind the total silliness of

Darwinian evolutionary hypotheses, it appears that "God" is at the top of the hierarchy of life forms and single-celled bacteria are positioned at the lowest level. Humans are near the top of the hierarchy beneath God and "angels." God, angels and humans are immortal spiritual beings existing as the trinity of body, soul and spirit. God is eternally present and functions as the sovereign within every physical and spiritual dimension. God created humans in His own image after His own likeness and incarnated Himself as Jesus Christ when He lived on Planet Earth as humanity's "sacrificial lamb." Jesus Christ was and forever is the express image of God. God has revealed Himself to humanity as "God, the Father, God, the Son, and God, the Holy Spirit."

Angels were created by God and gifted with a free will thereby having the freedom of choice to love, honor and obey God; or to hate, disobey and war against God and humans who were also gifted with a free will. Whether angel or human, the "spirit" within the individual is the life force enabling an immortal being to maintain "God awareness." The "soul" is the "mind" of the living being which exercises free will and is also the seat of emotions, psychic drives, and continuous decision making. The physical body is the molecular structure within which the spirit and soul experiences "world consciousness" and interacts with the dimensional environment.

The human body, soul and spirit were originally created by God to be immortal but

the first human couple (man and woman)
chose to defy God whereupon their spirits and
souls were banished from God's presence and
their physical bodies were appointed to death.
Spiritual banishment occurred simultaneously
with rebellion against God and physical death
followed from the divine curse of "aging"
whereby the human body returns to the
primordial elements from which it was created.

Humans were not alone in their rebellion
against God. A significant number of angels
chose to wage spiritual war against their
Creator long before humans were created and
given dominion of Planet Earth. Some of the
rebelling angels are chained in outer darkness
awaiting judgment. A significant number of
rebelling angels are being allowed by God to

temporarily wage spiritual warfare against Him and all human descendants of the man and woman (Adam and Eve) originally created by God. Such divine permission is for the purpose of allowing each individual human to freely choose an eternal spiritual father (God or Satan). By choosing Satan, Adam and Eve lost their physical immortality and consequently could only give birth to mortal offspring conceived in a state of rebellion and inherent iniquity.

Because God is the essence of pure divine love, He incarnated Himself and offered up His human body and blood through the humanity of Jesus Christ thereby becoming humanity's sacrificial lamb and redeeming fallen humans back to the eternal Godhead through a spiritual rebirth **upon belief in and acceptance** of Jesus Christ as their personal sacrificial lamb Who willingly paid the full cost of **human redemption** from sin, sorrow and eternal spiritual banishment into the "lake of fire prepared for Satan and his demonic angels."

The eternal Godhead sends no eternal spirit to hell and thereafter into the lake of fire following the "Great White Throne Judgment." The simple choice of **choosing** an eternal spiritual father **determines where immortal spiritual beings spend eternity.** Every immortal spirit that winds up in the lake of fire will be there because the immortal spirit **freely chose** to follow Satan into his eternal habitat.

The "Holy Scriptures" are the fully

documented and written record of God's
communications to humanity recorded by
more than three dozen human authors over a
period of sixteen centuries. The Holy Scriptures did not originate pursuant to the will of
humans; but rather authors selected by God
wrote down what He communicated to their
human spirit.

The Holy Scriptures were translated
from the original Hebrew and Greek to English
and hundreds of other languages and dialects
and canonized into a single book known as
"The Holy Bible" (hereinafter referred to as
"The Bible"). The accuracy of such translation
is demonstrated by the the total failure of
millions of attacks upon Biblical integrity
occurring continuously since the fourteenth

century prior to the human birth of Jesus
Christ. The "Old Testament" of The Bible was
written down between 1491 B.C. and 397 B.C.
The "New Testament" was written down
between 37 A.D. and 96 A.D.

Chapter four
Dispensation of Innocence

The Bible has survived every attempt to
eliminate the Holy Scriptures from the book
shelves of mankind. It is, without question, the
most incredible and most influential book ever
introduced into human cultures and has
actually given birth to Western civilization. It
is the most frequently translated and widely
read book in the history of the world. The
Bible has also been attacked more frequently
and more viciously than any other book ever
written. The most devout atheists and
evolutionists admit that the Bible is their most
formidable adversary and that Christianity is

actually growing exponentially in the twenty-first century after the Incarnate Word appeared on Earth. Nevertheless, popularity and longevity alone do not prove the Bible to be true, except in the sense that continued survival in the face of all out efforts to discredit its veracity indicates it can withstand the closest and most hostile scrutiny.

What is an indisputable test for determining whether the Bible is truth or fiction? Both the Old Testament and the New Testament writings have been in existence for over eighteen hundred years and have been under constant attack. Although the Biblical record was compiled over a period of sixteen centuries by thirty-seven different authors, the entire Bible reads like it was written by one individual and the central theme does not change from beginning to end. That central theme is Jesus Christ. In the Old Testament, he is the Messiah who will establish a kingdom that will endure forever. In the New Testament, he is the Lamb of God providing the supreme sacrifice for the sins of humanity, and the mediator between God and all humans who accept him as the divine sacrifice for their personal sins.

The New Testament proclaims Jesus Christ to be King of kings and Lord of lords; the Incarnate Word; Son of God and "seed of the woman;" alpha and omega, the beginning and the end; the eternal creator of everything in the universe; and the bodily manifestation of pure, undefiled love. There is no disagreement within Old Testament writings as to who Jesus Christ really is. In the eighth century B.C., before God incarnated himself in the body of Jesus Christ, It was written in the scroll of the prophet, Isaiah:

"For unto us a child is born, unto us a Son is given: and the government shall be upon His shoulder; and His name shall be called Wonderful, Counselor, Mighty God, the Everlasting Father, the Prince of Peace. Of the increase of His government and peace there shall be no end, upon the throne of David, and upon his kingdom, to order it, and to establish it with judgment and with justice from henceforth even forever. The zeal of the Lord of hosts will perform this."   (Isaiah, chapter 9, verses 6-7; 740 B.C. KJV)

It would be most reasonable then to focus our search for truth upon what the Old and New Testaments proclaim concerning Jesus Christ, whether as Messiah or Lamb of God. We will, however, look at statements and prophecies that address lesser themes. It is written that Jesus said: "I am Alpha and Omega, the beginning and the end, the first and the last."   (Revelation, chapter 22, verse 13, 96 A.D. KJV)

When was the beginning and when will the end be? The Bible does not contain a date for either the beginning or the end, and it gives more information concerning the end than it does about the beginning.   The Bible does state: "In the beginning God created the heaven and the earth." (Genesis, chapter 1,

verse 1; writer, Moses, 15th century B.C.
KJV). It is important to note that in this
statement, it is proclaimed that " in the
beginning" God created Earth and its
atmosphere (the heaven containing the clouds
of water vapor). But, in the very next verse, the
atmosphere is missing: "And the earth was
without form, and void; and darkness was
upon the face of the deep. And the Spirit of
God moved upon the face of the waters"
(Genesis, chapter 1, verse 2).

How much time elapsed between the
creation of Earth stated in verse 1 and the
condition of Earth pictured in verse 2? It
could have been a very long period of time, or
a brief period. Within the vastness of the
universe, time as measured by humans doesn't

appear to be a significant consideration.

Today's scientists tell us that it takes billions of years for the light coming from distant stars to reach Earth. Time is only important to those who know their life is brief. To an eternal being, the passing of time doesn't count for much. It is clear from the two statements that Earth suffered a devastating catastrophe and languished in a chaotic state for an undisclosed period of time. There are Scriptures which indicate that Earth had undergone a cataclysmic change as the result of divine judgment:

"I beheld the earth, and, lo, it was without form, and void: and the heavens, and they had no light." (Jeremiah, chapter 4, verse 23, vision 612 B.C. KJV)

"Behold, the Lord maketh the earth empty, and maketh it waste, and turneth it upside down, and scattereth abroad the inhabitants thereof." (Isaiah, chapter 24, verse 1, vision 715 B.C. KJV)

"For thus saith the Lord that created the heavens; God Himself that formed the earth and made it; He hath established it, He created it not in vain, He formed it to be inhabited: I am the Lord; and there is none else." (Isaiah, chapter 45, verse 18, 712 B.C. KJV)

"How art thou fallen from heaven, O Lucifer, son of the morning!.....For thou hadst said in thine heart, I will ascend into heaven, I will exalt my throne above the stars of God: I will sit also upon the mount of the congre-gation, in the sides of the north; I will ascend above the heights of the clouds; I will be like the Most High.(Isaiah, chapter 14, verses 12 to 14, vision 712 B.C. KJV)

".......Thus saith the Lord God; Thou sealest up the sum, full of wisdom, and perfect in beauty. Thou hadst been in Eden the garden of God; every precious stone was thy covering, the sardius, topaz, and diamond, the beryl, the onyx, and the jasper, the sapphire, the emerald, and the carbuncle, and gold: the workmanship of thy tabrets and of thy pipes was prepared in thee in the day that thou wast created. Thou art the anointed cherub that covereth; and I have set thee so; thou wast upon the holy mountain of God; thou hast walked up and down in the midst of the stones of fire. Thou wast perfect in thy ways from the day that thou wast

created, till iniquity was found in thee."
(Ezekiel, chapter 28, verses 12-15, vision 588
B.C. KJV)

"And there was war in heaven: Michael
and his angels fought against the dragon; and
the dragon fought and his angels, and prevailed
not; neither was their place found any more in
heaven. And the great dragon was cast out, that
old serpent called the Devil, and Satan, which
deceiveth the whole world: he was cast out
into the earth, and his angels were cast out
with him."   (Revelation, chapter 12, verses 7-
9, Patmos vision A.D. 96 KJV)

The composite of these passages
implicates Earth in the "fall of Lucifer," (also
called "Satan," "Devil," and "Dragon"). There
was an absence of energy lighting the

planet, and there was no division of liquids and gases on Earth's surface. Whatever living creatures inhabited Earth at that time perished as well as plant life. However, plant seed remained dormant within Earth. The physical shape of Earth reflected chaos such that it appeared "without form and void," and completely covered with water.

Genesis, chapter 1, verse 2 (KJV) begins the account as to how God brought order out of chaos, created humans and gave them dominion over the planet. The intense, eternal hatred Lucifer (Satan, Devil, Dragon) exhibits toward mankind is also understandable in the light of these composite Scriptures. As the first step in bringing order out of chaos, Genesis, chapter 1, verse 3 records:

"And God said, Let there be light: and there was light" (made to appear, made visible; the sun and moon were created "in the beginning" but some mass had invaded the solar system and blocked out the light radiating from the sun and reflecting from the moon).

A close examination of the actual recorded text contained within Genesis, chapters 1 and 2 reveals that the past tense (in the beginning) and the time period spanning the end of chaos and restoration of order upon Earth (preparing Earth for dominion by humans) are intertwined in order to maintain the continuity of the narrative in summary form. The Hebrew word for "made" translates into English as both "made" and "had made."

Keeping that fact in mind, the narrative in Genesis, chapters 1 and 2 makes perfect sense in its summary form. Chapter 1 presents a complete summary of "in the beginning" and "bringing order out of chaos." Chapter 2 adds specific details concerning the creation of humans within Earth prepared for dominion by mankind. Genesis, chapter 2, verses 7-9 explicitly state that mankind was created:

"And the Lord God formed man of the dust of the ground, and breathed into his nostrils the breath of life; and man became a living soul. And the Lord God planted a garden eastward in Eden; and there he put the man whom he had formed. And out of the ground made the Lord God to grow every tree that is pleasant to the sight, and good for food; the tree of life also in the midst of the garden, and the tree of knowledge of good and evil."

Although the Scriptures do not reveal the elapsed time spanning "in the beginning," nor how long Earth languished in a state of chaos, we can trace how long humans have inhabited the planet by the sum of the life spans of Adam and his descendants prior to the "days of Noah" (recorded in Genesis, chapter 5), plus the passing of time since Noah's death (grand total equal to roughly sixty centuries). Here again, since we have already demonstrated that Darwin's theory of evolution is only a figment of his imagination, there is no conflict between the verifiable scientific body of knowledge existing today and the Biblical account of how humans came into being. The

Scriptures do, however, answer many basic questions for which scientists have no other explanation whatsoever.

The fossilized remains of ancient, extinct life forms have been found, but there are absolutely no fossilized remains of humans more than sixty centuries old. When fossilized remains of life forms, other than humans, do appear in the fossil record, they are either extremely ancient or essentially the same as non-human life forms that inhabit Earth today, or have become extinct; and they all appear in the same time period.

It is patently obvious..... chemically, biologically, mathematically, and just using common sense that the universe and all matter therein could only have been designed and then created by God (an awesome creative power existing outside time and space). Every so-called "missing link" ever spaded up by disciples of Darwin has proven conclusively to be a brazen hoax or a simple case of mistaken identity.

The time window within which God created Eve demonstrates that humans were created within a single solar day:

"And the Lord God caused a deep sleep to fall upon Adam, and he slept; and he took one of his ribs, and closed up the flesh instead thereof; and the rib, which the Lord God had taken from man, made he a woman, and brought her unto the man. And Adam said, This is now bone of my bones, and flesh of my flesh: she shall be called Woman, because she was taken out of man.

Therefore shall a man leave his father
and his mother, and shall cleave unto his wife;
and they shall be one flesh. And they were
both naked, the man and his wife, and were not
ashamed. Now the serpent was more subtil
than any beast of the field which the Lord God
had made. And he said unto the woman, Yea,
hath God said, Ye shall not eat of every tree of
the garden? And the woman said unto the
serpent, We may eat of the fruit of the trees of
the garden: But of the fruit of the tree which is
in the mist of the garden, God hath said, Ye
shall not eat of it, neither shall you touch it,
lest ye die.

And the serpent said unto the woman, Ye
shall not surely die: For God doth know that in

the day ye eat thereof, then your eyes shall be opened, and ye shall be as gods, knowing good and evil. And when the woman saw that the tree was good for food, and that it was pleasant to the eyes, and a tree to be desired to make one wise, she took of the fruit thereof, and did eat, and gave also unto her husband with her; and he did eat. And the eyes of both of them were opened, and they knew that they were naked; and they sewed fig leaves together, and made themselves aprons." (Genesis, chapter 2, verse 21 through chapter 3, verse 7 KJV)

The forbidden tree was the only vehicle through which Adam and Eve could come to know the folly of sin and the loss of pure innocence. Moreover, they also knew the penalty for sin is death. Through sin against God they would become vulnerable to the living creatures opposing God (Lucifer and his angels). It is therefore most fitting that God called the forbidden tree the "tree of the knowledge of good and evil." The tree, itself, was of little significance. It was the act of willing disobedience coupled with the knowledge of the penalty that would "open their eyes" to the knowledge of good and evil and result in banishment from God's presence; immediate spiritual death followed by physical death. Hence, it was no small thing, this eating of the forbidden fruit.

Why did Adam and Eve, knowing the penalty, willingly break God's single commandment given to them? First, Eve

listened to Lucifer who is the father of lies:
".....Ye shall not surely die: For God doth
know that in the day ye eat thereof, then your
eyes shall be opened, and ye shall be as gods,
knowing good and evil." (Genesis, chapter 3,
verses 4-5)   Then, Eve allowed herself the
twin pleasures of lust and pride. The fruit was
pleasant to look at, it would probably be very
tasty, and it would make her as wise as God.
Besides, the serpent was more to be trusted
than her creator. God had been holding out on
her. Pride turned into arrogance. She reached
out and fondled death. It felt just like a ripe,
delicious fruit. Her breasts swelled with
rebellion and pride as she hurried to find Adam
to give him a taste. For the very first time, Eve
became aware of the force of human lust and

her own nakedness. Should Adam be less than enthusiastic about risking death, she could probably coax him into joining her.

Adam and Eve died spiritually that very day and were cast out of the garden of God. They were banished from God's presence to labor for their food and to conceive children in their own image of lust, pride and arrogance. They were now appointed to physical death and their children would be under the same death sentence. Immortal had become mortal, innocence had become lust, eternal life had been left behind in the garden of God.

Mortal living beings cannot pass on eternal life. Adam and Eve had sealed the fate of their children and all future generations. All would be born in a state of sin and mortality.

All human flesh was forever barred from God's eternal presence. But hope still existed for that breath of God imparted, through Adam and Eve, to every human in the form of an eternal living spirit. Even when casting Adam and Eve from the creation garden, God knew He would incarnate Himself and offer up His own body and blood to redeem human spirits and souls. Within the future of human flesh would be only physical, eternal death.

The incarnation of God into human flesh in the person of Jesus Christ is the central truth of the Bible and also is the acid test for Scriptural infallibility. If Jesus Christ is, in fact, who He claims to be, then the search for truth is over. Jesus not only said He is the son of God. He said he is God:

"I and my Father are one." (John, chapter 10, verse 30 KJV)   ".....he that hath seen me hath seen the Father...." (John, chapter 14, verse 9 KJV)

"In the beginning was the Word, and the Word was with God, and the Word was God" (John, chapter 1, verse 1 KJV).   ".....and the Word was made flesh and dwelt among us,......." (John, chapter 1, verse 14)

Centuries before Jesus appeared on Earth, He, Himself, and many details pertaining to his sacrificial death were described in astounding detail within Old Testament Scriptures:

"But thou, Bethlehem Ephratah, though thou be little among the thousands of Judah, yet out of thee shall He come forth unto me that is to be ruler in Israel; whose goings forth have been from old, from everlasting." (Micah, chapter 5, verse 2, 710 B.C. KJV).

"Therefore the Lord Himself shall give you a sign; Behold a virgin shall conceive, and bear a son, and shall call his name Immanuel." (Isaiah, chapter 7, verse 14, 742 B.C. KJV)

"Thus saith the Lord; A voice was heard in Ramah; lamentation, and bitter weeping; Rachel weeping for her children refused to be comforted for her children, because they were not." (Jeremiah, chapter 31, verse 15, 599 B.C. KJV)

"Rejoice greatly, O daughter of Zion; shout, O daughter of Jerusalem; behold, thy King cometh unto thee: He is just, and having salvation; lowly, and riding upon an ass, and upon a colt the foal of an ass." (Zechariah, chapter 9, verse 9, 487 B.C. KJV)

"Yea, Mine own familiar friend, in whom I trusted, which did eat of My bread, hath lifted up his heel against Me." (Psalm 41, verse 9, 11th century B.C. KJV)

".....He hath no form nor comeliness; and when we see Him, there is no beauty that we should desire Him. He is despised and rejected of men; a man of sorrows, and acquainted with grief: and we hid as it were our faces from Him; He was despised, and we esteemed Him not. Surely He hath borne our griefs, and carried our sorrows: yet we did esteem Him stricken, smitten of God and afflicted. But He was wounded for our transgressions, He was

bruised for our iniquities: the chastisement of
our peace was upon Him; and with His stripes
we are healed. All we like sheep have gone
astray; we have turned every one to his own
way; and the Lord hath laid on Him the
iniquity of us all. He was oppressed, and He
was afflicted, yet He opened not His mouth:
He is brought as a lamb to the slaughter, and as
a sheep before her shearers is dumb, so He
opened not His mouth. He was taken from
prison and from judgment: and who shall
declare His generation? for He was cut off out
of the land of the living: for the transgression
of my people was He stricken. And He made
His grave with the wicked, and with the rich in
His death; because He had done no violence,
neither was any deceit in His mouth....."

(Isaiah, chapter 53, verses 1-11, 712 B.C. KJV)

"I gave My back to the smiters, and My cheeks to them that plucked off the hair: I hid not My face from shame and spitting." (Isaiah, chapter 50, verse 6, 719 B.C. KJV)

".....and they shall look upon Me whom they pierced...." (Zechariah, chapter 12, verse 10, 487 B.C. KJV)

"And I said unto them, If ye think good, give Me my price; and if not, forbear. So they weighed for My price thirty pieces of silver. And the Lord said unto me, Cast it unto the potter: a goodly price that I was prised at of them. And I took the thirty pieces of silver, and cast them to the potter in the house of the Lord." (Zechariah, chapter 11, verses 12-13,

487 B.C. KJV)

"And one shall say unto Him, What are these wounds in Thy hands? Then He shall answer, Those with which I was wounded in the house of My friends." (Zechariah, chapter 13, verse 6, 487 B.C. KJV)

"My God, my God, why hast Thou forsaken Me?......All they that see Me laugh Me to scorn: they shoot out the lip, they shake the head, saying, He trusted on the Lord that He would deliver Him: let Him deliver Him, seeing He delighted in Him.....the assembly of the wicked have enclosed Me: they pierced My hands and My feet. I may tell all My bones: they look and stare upon Me. They part My garments among them, and cast lots upon My vesture."   (Psalm 22, verses 1-18,

11th century B.C. KJV)

"Reproach hath broken My heart; and I am full of heaviness; and I looked for some to take pity, but there were none; and for comforters; but I found none. They gave Me also gall for My meat; and in My thirst they gave Me vinegar to drink." (Psalm 69, verses 20-21, 11th century B.C. KJV)

Every minute detail of these prophetic passages was fulfilled in the birth, life and sacrificial death of Jesus Christ. The odds against this happening by sheer chance or by self-fulfilling prophecies are well beyond the realm of remote possibilities. This fact alone is sufficient to demonstrate absolute Biblical truth as well as incredible accuracy. It should be sufficient evidence to convince the most vehement skeptic that Jesus indeed came to bear witness to the truth. Of course, there is also the empty tomb and the hundreds of witnesses who personally interacted with Jesus after His resurrection and before He ascended back to the throne of Almighty God. There is nothing within the history of the universe better documented than the birth, life, death and resurrection of Jesus Christ. Those who reject him do so not for lack of Biblical and secular documentation, but rather for the same reasons Lucifer wanted to exalt himself above God Almighty....pride, arrogance and self-indulgence.

If sin is the willful rebellion against God's commandments, and if the law embodying the "ten commandments" was not

in effect until 1491 B.C., how can God hold
humans accountable for sin prior to 1491
B.C.? The answer is human "God consciousness," the "breath of God" which he breathed
into Adam's nostrils and which Adam passed
on to the human race. Every human knows the
difference between "good" and "evil." This
knowledge became part of human consciousness when Adam and Eve rebelled against God
in the Garden of Eden and opened the door to
eternal awareness of both good and evil. The
law written down by God and given by Moses
to the nation of Israel (the "seed of Abraham")
in 1491 B.C. set Israel apart from every other
nation on Earth when Israel, with one voice,
voluntarily accepted the law and bound
themselves with a blood covenant:

"And Moses took half of the blood, and put it in basins; and half of the blood he sprinkled on the alter. And he took the book of the covenant, and read in the audience of the people: and they said, All that the Lord hath said will we do, and be obedient. And Moses took the blood, and sprinkled it on the people, and said, Behold the blood of the covenant, which the Lord hath made with you concerning all these words." (Exodus, chapter 24, verses 6-8, 1491 B.C. KJV)

The law recognized the fallen state of humans and made provisions for the forgive- ness of sins through animal sacrifices which symbolized the future sacrifice God would offer up of His own body and blood in the person of Jesus Christ, the express image of

God. The nation of Israel was to be the vehicle
through which the Lamb of God would come
into the world as the Incarnate Word and offer
Himself for the sins of all humanity; and the
animal sacrifices contained in the law given to
Moses were symbolic of the divine sacrifice
"to be offered up in the fullness of time:"

"And it shall be, when he shall be
guilty.....that he shall confess that he hath
sinned.....And he shall bring his trespass
offering unto the Lord for his sin which he
hath sinned, a female from the flock, a lamb or
a kid of the goats, for a sin-offering; and the
priest shall make an atonement for him
concerning his sin. And if he be not able to
bring a lamb, then he shall bring for his
trespass, which he hath committed, two

turtledoves, or two young pigeons, unto the Lord; one for a sin offering, and the other for a burnt offering. And he shall bring them unto the priest, who shall offer that which is for the sin offering first, and wring off his head from his neck, but shall not divide it asunder: And he shall sprinkle of the blood of the sin offering upon the side of the alter; and the rest of the blood shall be wrung out at the bottom of the alter: it is a sin offering. And he shall offer the second for a burnt offering, according to the manner: and the priest shall make an atonement for him for his sin which he hath sinned, and it shall be forgiven him." (Leviticus, chapter 5, verses 5-10, 1491 B.C. KJV)

"And in thy seed shall all the nations of the earth be blessed; because thou hast obeyed

my voice." (Genesis, chapter 22, verse 18,
1872 B.C. KJV)

The seed of Abraham (Israel) was to
evangelize all nations and teach them God's
plan of redemption. Until Israel accomplished
this mission, God would continue to deal with
individuals who sought him out in accordance
with the "God consciousness" which God
imparted to every human through Adam and
Eve. There has never been individual
ignorance of God's existence, nor will there
ever be:

"For the wrath of God is revealed from
heaven against all ungodliness and unrighteousness of men, who hold the truth in
unrighteousness; Because that which may be
known of God is manifest in them; for God

hath shewed it unto them. For the invisible things of Him from the creation of the world are clearly seen, being understood by the things that are made, even His eternal power and Godhead; so that they are without excuse. Because that, when they knew God, they glorified Him not as God, neither were thankful; but became vain in their imagin-ations, and their foolish heart was darkened. Professing themselves to be wise, they became fools, and changed the glory of the incorrupt-ible God into an image made like to corrupt-ible man, and to birds, and four-footed beasts, and creeping things.

Wherefore God also gave them up to uncleanness through the lusts of their own hearts, to dishonor their own bodies between

themselves: Who changed the truth of God
into a lie, and worshiped and served the
creature more than the Creator, who is blessed
forever, Amen. For this cause, God gave
them up unto vile affections: for even their
women did change the natural use into that
which is against nature:

And likewise also the men, leaving the
natural use of the woman, burned in their lust
one toward another; men with men working
that which is unseemly, and receiving in
themselves that recompense of their error
which was meet. And even as they did not like
to retain God in their knowledge, God gave
them over to a reprobate mind, to do those
things which are not convenient; Being filled
with all unrighteousness, fornication, wicked-

ness, covetousness, maliciousness; full of
envy, murder, debate, deceit, malignity,
whisperers, backbiters, haters of God, despite-
ful, proud, boasters, inventors of evil things,
disobedient to parents, without understanding,
covenant-breakers, without natural affection,
implacable, unmerciful: Who knowing the
judgment of God, that they which commit such
things are worthy of death, not only do the
same, but have pleasure in them that do them."
(Romans, chapter 1, verses 18-32, 60 A.D.
KJV)

In view of this communication, directly
from Almighty God through the inspired
writing of Paul, the Apostle, it would be highly
inadvisable for anyone in the day of judgment
to rely on ignorance as an excuse. Here, in this

passage, God plainly states that ignorance is the defense of liars. God also makes it very, very clear that homosexuality is not the result of confused sexuality in the womb, or any other cause beyond the control of the individual. Rather, it is the direct result of a personal decision to flaunt free will and personal rebellion against God. In the day of judgment, there are no excuses God will hear:

"And I saw the dead, small and great, stand before God; and the books were opened: and another book was opened, which is the book of life: and the dead were judged out of those things which were written in the books, according to their works." (Revelation, chapter 20, verse 12, A.D. 96 KJV).

The truly marvelous aspect of God's

sacrifice of Himself to redeem human souls is that all who accept Jesus Christ as their personal redeemer will not be present at the judgment described in Revelation, chapter 20, verse 12:

"As far as the east is from the west, so far hath He removed our   transgressions from us." (Psalm 103, verse 12, 11th century B.C. KJV)

"Come now, and let us reason together, saith the Lord: though your sins be as scarlet, they shall be white as snow; though they be red like crimson, they shall be as wool." (Isaiah, chapter 1, verse 18, 760 B.C. KJV)

"For God so loved the world, that He gave His only begotten Son, that whosoever believeth in Him should not perish, but have everlasting life. For God sent not His Son into the world to condemn the world, but that the world through Him might be saved." (John, chapter 3, verses 16-17, A.D. 90 KJV)

There is no lack of knowledge of God in the world; it is the lack of respect and worship that condemns God haters. Billions of humans would rather pray to a graven image made of wood, stone, or plastic because they can satisfy the lust of the flesh, the lust of the eyes and the pride of life; then kneel before an idol and ask for health and strength to keep doing the same things. God is both merciful and just; unforgiven sin, however, He will never tolerate. Do not plan on ignorance as an excuse.....says God.

God knows and understands the depth of

temptations humans face every day: "For we have not a high priest which cannot be touched with the feelings of our infirmities; but was in all points tempted like as we are, yet without sin." (Hebrews, chapter 4, verse 15, 64 A.D. KJV)

"There hath no temptation taken you but such as is common to man: but God is faithful, who will not suffer you to be tempted above that ye are able; but will with the temptation also make a way of escape, that ye may be able to bear it." (1st Corinthians, chapter 10, verse 13, 59 A.D. KJV)

God's holiness and justice demand that the sacrifice for our sins must be without sin and be subjected to the same temptations as mankind. Thus, God incarnated Himself in human form, subject to human needs, limitations, and temptations. Jesus Christ hungered, thirsted, grew tired, needed sleep, needed protection from the elements, and was mightily tempted by Lucifer. He felt sorrow, rejection, and anger. He was moved with compassion and pity. He wept and he rejoiced; but He did not sin. He was not begotten by a man and therefore was not born in a sinful state. He was the "seed of the woman" (a virgin) overshadowed by the Spirit of God. He was the one and only begotten son of God, and was still God, Himself. He lived without sin and then became sin in our place thereby setting us free from sin and the penalty therefor. The first prophesy concerning the redemption of humanity by Jesus Christ was uttered by God in the Garden of Eden:

"And I will put enmity between thee and the woman, and between thy seed and her seed; It shall bruise thy head, and thou shalt bruise His heel." (Genesis, chapter 3, verse 15 KJV)

In the process of redeeming mankind, Satan would be allowed to bruise Jesus, but Jesus would take away all of Satan's powers......forever.

Chapter five
Human Government

The spiritual war between the angels who remained faithful to God and the angels who joined Satan's rebellion had been raging for an undisclosed duration before order was restored upon Planet Earth and dominion of Earth given by God to Adam and Eve.

Although the Bible does not provide specific details, the initial judgment of fallen angels left Planet Earth "without form and void" and submerged in water. A passage of Scripture found in the scroll of the prophet Isaiah indicates that Lucifer (Satan) had his angelic throne upon Earth prior to his rebellion against God:

""How art thou fallen from heaven, O Lucifer, son of the morning!.....For thou hadst said in thine heart, **I will ascend into heaven, I will exalt my throne above the stars of God:** I will sit also upon the mount of the congregation, in the sides of the north; I will ascend above the heights of the clouds; I will be like the Most High. (Isaiah, chapter 14, verses 12 to 14, vision 712 B.C. KJV)

It is noteworthy that Lucifer desired to ascend into heaven thus indicating that he was somewhere below God's throne. He also wanted to sit within "the sides of the north" and be "like the Most High." His deadly and lethal opposition to humans also supports the assumption that Earth was formerly his domain. Through his success in deceiving Eve, he stole the undisputed dominion of Earth which God had given to humans. By joining Satan's rebellion, Adam and Eve adopted Satan as their spiritual father prior to their expulsion from the Garden of Eden. In order to be redeemed from their fallen state, their immortal spirits banished from God's presence would have to be "reborn."

"There was a man of the Pharisees, named Nicodemus, a ruler of the Jews: The same came to Jesus by night, and said unto him, Rabbi, we know that Thou art a teacher come from God: for no man can do these miracles that Thou doest, except God be with him. Jesus answered and said unto him, Verily, verily, I say unto thee, Except a man be born again, he cannot see the kingdom of God.

Nicodemus saith unto Him, How can a man be

born when he is old? Can he enter the second

time into his mother's womb, and be born?

Jesus answered, Verily, verily, I say unto thee,

Except a man be born of water and of the

spirit, he cannot enter into the kingdom of

God. That which is born of the flesh is flesh;

and that which is born of the spirit is spirit.

Marvel not that I said unto thee, Ye must be

born again. The wind bloweth where it listeth,

and thou hearest the sound thereof, but canst

not tell whence it cometh, and whither it goeth:

so is everyone that is born of the spirit.

Nicodemus answered and said unto Him, How

can these things be? Jesus answered and said

unto him, Art thou a master of Israel, and

knowest not these things? Verily, verily, I say
unto thee, We speak that We do know, and
testify that We have seen; and ye receive not
Our witness. If I have told you earthly things,
and ye believe not, how shall ye believe, if I
tell you of heavenly things? And no man hath
ascended up to heaven, but He that came
down from heaven, even the Son of man which
is in heaven. And as Moses lifted up the
serpent in the wilderness, even so must the Son
of man be lifted up: That whosoever believeth
in Him should not perish, but have eternal life.
For God so loved the world, that He gave His
only begotten Son, that whosoever believeth in
Him should not perish, but have everlasting
life. For God sent not His Son into the world
to condemn the world; but that the world

through Him might be saved." (John, chapter 3, verses 1-17, Jesus speaking, 30 A.D. KJV)

In accordance with eternal foreknow- ledge, God ever foresaw the cross upon which He would, through the willing humanity of Jesus Christ, offer up His body and blood to redeem fallen humanity. Consequently, the Old Testament of The Bible looks forward to the birth, ministry, crucifixion and resurrection of Jesus; and the New Testament looks back at the "cross of redemption."

Every human from Adam and Eve to the present time, through "God consciousness" and the "knowledge of good and evil," knew instinctively what human behavior was and is acceptable to God and what human behavior constitutes iniquity (evil acts whether or not expressly forbidden by God). Prior to the actual crucifixion of Jesus Christ, human acts constituting iniquity were temporarily forgiven by offering up the body and blood of an innocent sacrificial animal which symbolically pointed to the sacrificial death of Jesus "in the fullness of time."

Although the spirits and souls of Adam and Eve and their offspring were banished from God's presence until the physical death of their human bodies, prayers of confession and repentance could be offered up to God along with the body and blood of the "sacrificial animal." Before driving Adam and Eve out of the Garden of Eden, God killed innocent animals and clothed Adam's and Eve's naked bodies with the bloody animal skins.

God had created vegetation suitable for human consumption before creating Adam and Eve and the Garden of Eden:

"And God said, Let the earth bring forth grass, the herb yielding seed, and the fruit tree yielding fruit after his kind, whose seed is in itself, upon the earth: and it was so. And the earth brought forth grass, and herb yielding seed after his kind, and the tree yielding fruit, whose seed was in itself, after his kind: and God saw that it was good. And the evening and the morning were the third day." (Genesis, chapter 1, verses 11 -12 KJV)

"These are the generations of the heavens and of the earth when they were created, in the day that the Lord God made the earth and the heavens, and every plant of the field before it was in the earth, and every herb of the field before it grew: for the Lord God had not caused it to rain upon the earth, and there was not a man to till the ground. But there went up a mist from the earth, and watered the whole face of the ground." (Genesis, chapter 2, verses 4-6 KJV)

Adam and Eve tilled the soils of earth and established flocks of animals for meat and clothing while mating with each other and bringing forth sons and daughters. Their first born son (Cain) murdered his younger brother (Able) during a dispute over whether to offer God the body and blood of a sacrificial animal versus the "fruit of the ground." Cain fled from Adam and Eve and took along a sister as his wife. Adam and Eve produced a son following the flight of Cain and named this third son Seth. The descendants of Adam and Eve and the descendants of Cain remained separate and hostile as the human population of Planet Earth increased geometrically over a period of sixteen centuries.

A small remnant of Seth's descendants worshiped God through prayer and sacrifice but Cain's descendants continued in open rebellion against God. Gradually, the descendants of Cain managed to convert the descendants of Seth until only one man (Noah) and his wife plus his three sons and their wives followed after God. As an unspeakable act of mercy, God elected to repopulate the earth with Noah's descendants thereby avoiding the absolute corruption of humanity under Satan's evil power and influence which corrupted the entire creative structure of Earth and its inhabiting life forms.

"And God saw that the wickedness of man was great in the earth, and that every imagination of the thoughts of his heart was only evil continually. And it repented the Lord that He had made man on the earth, and it grieved Him at His heart. And the Lord said, I will destroy man whom I have created from the face of the earth; both man, and beast, and the creeping thing, and the fowls of the air; for it repenteth Me that I have made them. But Noah found grace in the eyes of the Lord." (Genesis, chapter 6, verses 5-8 KJV)

God gave Noah detailed instruction for building a huge three-story ark designed to float upon turbulent waters. The ark was 450 feet long, 75 feet wide and 45 feet high thereby enclosing one million, five hundred and eighteen thousand, seven hundred and fifty cubit feet of interior space which is equal to one hundred and fifty-one railroad boxcars with 10,000 cubit feet per boxcar.

God sent a male and female of every beast, every fowl of the air, and every creeping thing to Noah to preserve them inside the ark; along with food supplies for everything that entered into the ark which included seven of each animal suitable for sacrificial offerings.

The global flood which extended above the highest mountains on Earth did not abate for one hundred and fifty days. Noah and the inhabitants of the ark exited to dry ground ten and one half months after the beginning of the global flood.

"And God blessed Noah and his sons, and said unto them, Be fruitful, and multiply, and replenish the earth. And the fear and the dread of you shall be upon every beast of the field, and upon every fowl of the air, and upon all that moveth upon the earth, and upon all the fishes of the sea; into your hand are they delivered. Every moving thing that liveth shall be meat for you; even as the green herb have I given you all things. But flesh with the life thereof, which is the blood thereof, shall ye not eat. And surely your blood of your lives will I require; at the hand of every beast will I require it, and at the hand of man; at the hand of every man's brother will I require the life of man. Whoso sheddeth man's blood, by man shall his blood be shed: for in the image of God made He man." (Genesis, chapter 9, verses 1-6 KJV)

"And the sons of Noah, that went forth of the ark, were Shem, and Ham, and Japheth: and Ham is the father of Canaan. These are the three sons of Noah: and of them was the whole earth overspread." (Genesis, chapter 9, verses 18-19 KJV)

Between the time window spanning the centuries from the global flood to the exodus of the Hebrew slaves from Egypt, human government was carried out by "heads of households," tribal chiefs, Dukes, Sultans, Sheikhs, Pharaohs, Princes, Kings, Emperors, and other dignitaries wielding power by placating the mass of the people being governed. Within cultural boundaries civil, criminal, and religious laws, statutes and ordinances were generally published in some form comprehensible to the common people. Idol worship involving multiple imaginary gods and goddesses and embracing human sacrifice, religious prostitutes, sodomy, genocide, homosexuality, slavery, and every other form of human depravity flourished under Satan's influence as Prince of the Power of the Air and God of the Earth exercising the total dominion of Earth yielded up to him by Adam and Eve. From the days of Noah until the present time, the vast majority of humanity have continued to freely choose Satan as their eternal spiritual father.

Being neither a liar nor "Indian Giver," God allows Satan to temporarily exercise dominion over Earth while also allowing each human the freedom to worship Satan:

"I call heaven and earth to record this day against you, that I have set before you life and death, blessing and cursing: therefore choose life, that both thou and thy seed may live." (Exodus, chapter 30, verse 19 KJV)

"Be not deceived; God is not mocked: for whatsoever a man soweth, that shall he also reap. For he that soweth to his flesh shall of the flesh reap corruption; but he that soweth to the spirit shall of the spirit reap life ever-lasting." (Galatians, chapter 6, verses 7-8 KJV)

It is foolish indeed to sow lettuce and

expect to reap tomatoes; or to sow sodomy, hate, adultery and murder and expect to reap love, forgiveness, redemption and eternal life in the Kingdom of God. This truth concerning sowing and reaping is forever embedded into the conscience of every human along with the knowledge of behavior triggered by iniquity.

For example, Pharaoh Ramses sowed the death of all of Egypt's first born children when he decreed that every male child born to Hebrew slaves must be fed to the Nile River crocodiles. Pontius Pilate sowed his own suicide when he crucified an innocent man to please religious hypocrites threatening rebellion against his rule. King Herod sowed being eaten by worms when he ordered the murder of Jewish children in Bethlehem. King

David sowed the bloody sword wielded against his own family when he murdered Uriah so that he could enjoy sex with Bathsheba. Haman sowed his own execution by hanging when he plotted to hang Ester's uncle. Every human who winds up in the lake of fire sowed this eternal destiny when adopting Satan as eternal spiritual father.

In the fullness of prophetic time as God had revealed to the Hebrew prophets, God chose the human seed through which Jesus Christ would be born into the world of mankind. Approximately 1975 B. C. God selected a man named Abraham for special blessings because Abraham believed God existed and wanted to please Him. God promised Abraham that he would father many

nations and instructed him to leave his
homeland and travel to a country which God
would give to him and to his seed forever.
Abraham gathered his household and journey-
ed from Ur of the Chaldees to Canaan where
he became a shepherd and waited for the
fulfillment of God's promises. Because Sarah,
his wife, was barren, Abraham took her advice
and fathered a son (Ishmael) by her handmaid.
God rejected Ishmael as Abraham's heir and
told Abraham that Sarah would bear him a son
with whom God would honor his promises to
Abraham. Abraham begat Isaac when he was a
hundred years old and when Sarah was far
beyond the age of childbearing.

God selected circumcision as the sign of
the covenant He made with Abraham and his

seed after him.

God told Abraham to offer his son, Isaac, as a sacrificial offering and then provided a substitute ram when Abraham actually raised his knife over Isaac as Isaac lay bound upon the altar (a preview for Abraham of the sacrifice of God, the Son). After the death of Sarah, Isaac took his cousin, Rebecca, as his wife and she bare twin sons named Esau and Jacob. Esau exited the womb first but sold his firstborn birthright to Jacob for a bowl of pottage to assuage his temporary hunger. Jacob thus inherited the covenant promises God made to Abraham; and Esau's descendants are the Edomites.

God changed Jacob's name to Israel and he begat twelve sons who are the twelve patriarchs of the "twelve tribes of Israel" (the tribes of Reuben, Judah, Gad, Asher, Simeon, Levi, Dan, Naphtali, Issachar, Zebulun, Joseph, and Benjamin).

The Children of Israel were taken captive into Assyria and Babylon pursuant to sowing idolatry, homosexuality and Satan worship. The Israelites have been either directly ruled or been subject to Gentile overlordship for more than twenty-five centuries. Around 588 B.C. a Jewish remnant returned from Babylon to resettle Jerusalem and rebuilt the Jewish Temple. Approximately 483 years after said remnant returned to Jerusalem, Jesus Christ was born of a virgin of the seed of King David during the occupation of Jerusalem by the Romans according to the Hebrew prophecies.

Chapter six
Law Covenant

The Hebrews spent four centuries as slaves to the Egyptians. During the lifetime of Jacob, the descendants of Abraham settled in Egypt following a seven year famine in the Middle East. The Hebrews prospered immensely and their offspring became exceedingly numerous. The Egyptians feared that the seed of Abraham, Isaac and Jacob might join the enemies of Egypt and dispossess the Egyptians. Therefore Pharaoh Ramses decreed that the Hebrew children born as females should serve the Egyptians as sex slaves and Hebrew children born as males should be fed to the Nile River crocodiles.

Moses was born to a Hebrew mother who hid him from the Egyptians for three months after which she set him afloat in the Nile River inside a basket thereby trusting Moses' future to God. Pharaoh's daughter saw the basket and found Moses inside. She adopted Moses as her own son and raised him in Pharaoh's palace.

Moses became aware of his Hebrew heritage from his natural mother who was unknowingly hired by Pharaoh's daughter to nurse him for wages. At the age of forty, Moses killed an Egyptian for mercilessly beating a Hebrew slave and then fled from Pharaoh to the Sinai wilderness for forty years to escape punishment for murder.

Moses protected the daughters of Jethro,

the Priest of Midian, from desert shepherds
who bullied the women at flock watering
wells. Jethro gave his daughter, Zipporah, as
wife to Moses and she bore Moses two sons,
Gershom and Eliezer. God spoke to Moses
from the midst of a burning bush which the
fire did not consume and instructed Moses to
return to Egypt and lead the Hebrews out of
slavery pursuant to a series of devastating
plagues which God would send upon Egypt
when Pharaoh refused to free the Hebrews.

Pharaoh hardened his heart through
plagues involving putrid waters, frogs, lice,
swarms of flies and locusts, diseased livestock,
hail and fire storms, grievous flesh sores, thick
darkness, and finally the death of all the first
born sons and daughters of the Egyptians after

which Pharaoh freed the Hebrews sending them out laden with the treasures of Egypt.

The night that the death angel killed the Egyptian first born, God told Moses to have the Hebrews, by families, kill a sacrificial lamb and swab the blood, in the form of a cross, over the door of their dwellings thus allowing the death plague to "pass over" the Hebrews. The Jewish "Feast of Passover" celebrates this "passing over."

However, after committing his own dead son to his imaginary gods, Pharaoh elected to pursue after the Hebrews and slaughter them. God divided the waters of the Red Sea thus allowing the Hebrews to cross the sea by foot into the Sinai wilderness. Pharaoh and his army foolishly attempted to follow after the escaping slaves whereupon God drowned the Egyptians in the depth of the sea by returning the waters to their natural balance.

The Hebrew exodus from Egyptian slavery occurred 1,491 years prior to the birth of Jesus Christ. God's plan for the descendants of Abraham embraced testimony to the one true God in the midst of universal idolatry; exhibiting to the world of humanity the blessedness of serving the God of creation; receiving, preserving and making available the thoughts and words of God directed to all nations, kindreds and tongues; and to serve as the human seed through which a virgin would conceive by the overshadowing of God's Holy Spirit thereby permitting Jesus Christ to be born as a human being apart from Adam's sin.

The Hebrews having been brutalized by Satan in Egypt became a stiff-necked, stubborn and rebellious people. They sang praises to God when crossing the Red Sea on dry ground and then proclaimed they were capable of doing whatever God instructed them to do pursuant to their own personal righteousness. They reasoned they would be entitled to God's ongoing blessings because of their personal efforts and good behavior.

To demonstrate to the Hebrews their evil nature, entirely beyond their personal ability to control their Satanic impulses, God called Moses up into the heights of Mount Sinai and gave him the criminal, civil, sacrificial and ceremonial laws summarized by the "Ten Commandments." The commandments simplified 613 individual statutes, judgments, and ordinances which could all be fulfilled by honoring God with mind, soul and body; and loving other humans as one loves his/her own self. Because the law given to Moses equates evil thoughts with evil acts, no human other than Jesus Christ has ever been able to keep such laws. God's purpose in giving the laws to the Hebrews was to make them fully aware that they needed a "divine sacrificial lamb" to trigger a "rebirth of their human spirits" thus making possible a new choice of eternal spiritual father (God instead of Satan).

Such is precisely what God through the Apostle Paul tells us in his epistle to the Christian Church in Galatia whose member-ship were being taught that we are redeemed from the slavery of sin back to our Heavenly Father by a combination of faith in Jesus Christ **plus** keeping the laws God gave to Moses:

"A man is not justified by the works of the law, but by exercising faith in Jesus Christ. We are justified before God by our faith in Jesus Christ and not by keeping the law: for by keeping the law shall no flesh be justified ...... I do not frustrate the Grace of God. If redemption comes by law keeping, then Christ is dead in vain." (Galatians, chapter 2, verses 15, 16, and 21 KJV with minor paraphrasing for clarity by author)

"Legalists" (those who teach salvation from sin through faith in Christ plus keeping the laws given to Moses) are absolutely incapable of keeping the laws they preach to others and those preached to are also incapable of keeping the laws through sheer will power. At first glance the laws summed up within The Ten Commandments appears easy to adhere to through human effort while exercising free will. The problem is that violation of God's moral, social and sacrificial laws occurs in our minds **first** whether or not actually acted out. Jesus Christ made this very plain when teaching those willing to listen to Him:

"You have heard that it was said by them of old time, Thou shalt not commit adultery: But I say unto you, that whosoever looketh on a woman to lust after her hath committed adultery with her already in his heart." (Matthew, chapter 5, verses 27-28 KJV).

All sin (violation of any one of God's laws) occurs first within our thoughts and then is visualized in our innermost being allowing sinful pleasure to be experienced separate from the overt act. Whatever is more important to us than God's love and forgiveness through faith in Jesus Christ is **spiritual idolatry.** Jesus pointed this out during His discourse with a man who trusted in his wealth. The man was a legalist and relied upon his keeping of God's Ten Commandments as his ticket to eternal life within the Kingdom of God. Jesus' response to him is admonitory:

"And when He was gone forth into the way, there came one running, and kneeled to Him, and asked Him, Good Master, what shall I do that I may inherit eternal life? And Jesus said unto him, Why callest thou Me good? There is none good but one, that is, God. Thou knowest the commandments, Do not commit adultery, Do not kill, Do not steal, Do not bear false witness, Defraud not, Honor thy father and mother. And he answered and said unto Him, Master, all these have I observed from my youth. Then Jesus beholding him loved him, and said unto him, One thing thou lackest: go thy way, sell whatsoever thou hast, and give to the poor, and thou shalt have treasure in heaven: and come, take up thy cross, and follow Me. And he was sad at that saying, and went away grieved: for he had great possessions." (Mark, chapter 10, verses 17-22 KJV)

The rich man's real problem was his sin

nature genetically inherited from Adam and Eve. Their life forces (spirits) has been banished from God's presence and Satan had become their spiritual father. All of the human race descended from Adam and Eve were and are conceived and born in sin with a rebellious spirit steeped in enmity toward God and totally incapable of a spiritual rebirth through our own efforts and exercise of free will.

Unredeemed humans are sinners not because of what they choose to do but rather because of their sin nature. A dog behaves like a dog. A pig behaves like a pig and sinners behave like a sinner because that is their inherited nature. A spiritual rebirth is only possible by a sovereign act of God at the precise time that a human believes in and accepts Jesus Christ as his/her personal sacrificial lamb to redeem his/her spirit back to God. Until being "born again" every human is stuck with Satan as spiritual father.

Considering that neither unbelievers nor believers in Jesus Christ are capable of keeping God's laws while their souls and spirits are confined within physical bodies appointed by God to physical death, the only and limited purpose of the laws was to expose the inbred sinful nature of humanity thereby pointing humans to Jesus Christ for redemption back to God. Where there is no law, there is no sin because by God's definition sin is the transgression of His laws. The laws given to Moses were therefore a perfect mirror to reflect human lust, greed, depravity and self-centered existence. Prior to spiritual rebirth and indwelling of God's Holy Spirit, no human is able to control the lust of the flesh, the lust of the eyes and the pride of life. God's laws were given solely to demonstrate conclusively to humanity that nothing but a divine sacrificial lamb could save mankind from eternal banishment from God for having freely chosen Satan as eternal spiritual father.

The laws were God's chosen method of bringing humans to the end of self-effort and to total reliance upon Jesus Christ for redemption. Jesus completely fulfilled God's laws on behalf of humanity and was therefore "the end of the law" for all who were willing to believe in Him and to accept Him as their personal sacrificial lamb. When the first created humans rebelled against God in their garden paradise, they were cast out under sentence of physical death into a hostile environment ruled by Satan to whom they had yielded up the dominion of Earth. They were banished from God's presence but carried with them the "breath of God" (their life force). Unfortunately, for humanity, they also carried with them the spirit of Satan whom they embraced in the garden. Thus, from birth to physical death, every human must endure an internal war between the breath of God and the spirit of Satan. Paul, the Apostle expressed this life long struggle:

"For that which I do I allow not: for what I would, that do I not; but what I hate, that do I. If then I do that which I would not, I consent unto the law that it is good, now then it is no

more I that do it but sin that dwelleth in me.
For I know that in me (that is in my flesh)
dwells no good thing: for to will is present
with me; but how to perform that which is
good I find not. For the good that I would I do
not: but the evil which I would not, that I do.
Now if I do that I would not, it is no more I
that do it, but sin that dwells in me. I find then
a law, that when I would do good, evil is
present with me. For I delight in the law of
God after the inward man: but I see another
law in my members, warring against the law of
my mind, and bringing me into captivity to the
law of sin in my members. O wretched man
that I am! Who shall deliver me from the body
of this death? I thank God through Jesus Christ
our Lord. So then with the mind I myself serve

the law of God; but with the flesh the law of sin." (Romans, chapter 7, verses 15-24 KJV)

God communicates with our spirit and Satan seeks to overwhelm our spirit with the lusts and desires of our flesh which has been condemned to physical death. We cannot through the exercise of our free will overcome the lusts and desires of our flesh because Satan resides in our condemned flesh.

Without a sacrificial lamb to redeem us back to God and give us victory over Satan, we are lost, alienated from God and without hope of redemption. For this reason, God sent Jesus Christ to pay the penalty for our sins and redeem us back to a state of innocence. When we accept Jesus Christ as our sacrificial lamb, God imparts to us His Holy Spirit by which we

defeat Satan in our condemned flesh. We are
"born again" and strive to do those things we
know are pleasing to God who has, through the
sacrifice of Jesus Christ, been restored to His
rightful status as our "Heavenly Father."

We will not be perfect in this mortal life
being still confined in our sinful and condemned flesh but our orientation and our desire will
be that of **son to Father**. We seek to please
God not in order **to be** redeemed but rather
because we **are** redeemed we love and
reverence God, our Heavenly Father.

The Old Testament animal sacrifices
offered up in accordance with the law of
Moses were a prophetic rendering of the
eternal sacrifice which God offered up of
Himself in the fullness of time. All redeemed

humans from the creation of man to the end of
measured time were and are forgiven by
mentally accepting the sacrifice which God
provided to pay the penalty for their sins. All
the sacrifices offered under the law of Moses
simply foreshadowed the coming of Jesus
Christ. We are redeemed today by believing
the record God gave us of the ministry,
sacrificial death and subsequent resurrection of
His Son, Jesus Christ.

Old Testament people were redeemed
back to God by believing in Him, trusting in
His word delivered by the prophets, and by
offering up blood sacrifices which were a
substitute for the future sacrificial death and
subsequent resurrection of God, the Son in the
person of Jesus Christ. Not only is the shed

blood of Jesus Christ the exclusive price paid to redeem humanity, the price of redemption was fully determined before the creation of mankind:

"Forasmuch as you know that you were not redeemed with corruptible things, as silver and gold, from your vain conversation received by tradition from your fathers; but with the precious blood of Christ, as of a lamb without blemish and without spot: who was foreordained before the foundation of the world, but was manifest in these last times for you. Who by Him do believe in God, that raised Him up from the dead, and gave Him glory; that your faith and hope might be in God." (1st Peter, chapter 1, verses 18-21 KJV)

God, being an all powerful, all knowing

and creative spirit, comprehends the past, present and future simultaneously. Thus, God was not surprised by human disobedience (sin) in the beginning of measured time. God had foreordained the price of human redemption and Jesus Christ was willing to pay the price. The only part humans play in the divine plan of redemption (salvation) is the belief in and acceptance of the sacrifice which God provided of Himself in the person of Jesus Christ.

Therefore, during unmeasured eternity, God will have exactly what He envisioned when He created humans.....living beings in His Own Image who love, reverence and fellowship with Him because they choose to do so. Paul, the Apostle explains the logic and justice of human redemption through our sacrificial lamb (Jesus Christ):

"Wherefore, as by one man sin entered into the world, and death by sin; and so death passed upon all men for that all have sinned: For until the law (law of Moses) sin was in the world: but sin is not imputed when there is no law. Nevertheless death reigned from Adam to Moses, even over them that had not sinned after the similitude of Adam's transgression, who is the figure of Him that was to come. But not as the offense, so also is the free gift. For if through the offense of one many be dead, much more the grace of God, and the gift by grace, which is by one man, Jesus Christ, has abounded unto many. And not as it was by one that sinned, so is the gift: for the judgment was by one to condemnation, but the free gift is of many offenses into justification. For if by one man's offense death reigned by one; much more they which receive abundance of grace and of the gift of righteousness shall reign in life by one, Jesus Christ. Therefore, as by the offense of one judgment came upon all men to condemnation; even so by the righteousness of one the free gift came upon all men unto justification of life. For as by one man's disobedience many were made sinners, so by the obedience of one shall many be made righteous. Moreover, the law entered, that the offense might abound (humanity aware of their sinning against God); but where sin abounded, grace did much more abound. That as sin has reigned onto death, even so might grace reign through righteousness unto eternal life by Jesus Christ our Lord." (Romans, chapter 5, verses 12-21 KJV)

In this single passage is summed up the exact purpose and content of God's word penned down in the Holy Bible. There is no excuse for any human to reject God's free gift of eternal life in His kingdom because God requires nothing but individual acceptance of the gift of redemption and subsequent eternal life in fellowship with the Godhead as opposed to an eternal existence in "the lake of fire prepared for Satan and his followers."

Every human since Adam has followed either God or Satan through the exercise of his/her free will. The good works that humble Christians strive for are not for the purpose of redemption but out of love, respect, adoration, and the fervent desire to be pleasant in the sight of our Heavenly Father.

All those who join Satan in his eternal punishment will be doing so by individual choice. Redemption is totally free whereas heavenly rewards are given out for overcoming the lust of the eyes, the lust of the flesh and the pride of life (often described as "faithful stewardship"). Contrary to religious beliefs and traditions around the world, there has always been and will always be only **one plan** of redemption for humans ordained by God.

Jesus Christ remains the central focus of the Holy Scriptures. The sacrificial offerings commanded by God for the temporary covering of sin mirrored the sacrifice that

Jesus would make of Himself as the one time and final sacrifice for the sins of humanity, past, present and future.

Individual human either accept Jesus Christ as their personal sacrificial lamb or they do no accept Him as such. Those who accept Him are forgiven forever by God and the laws given by Moses no longer apply to them. Therefore, sin is not imputed to them because sin is the transgression of the laws given to Moses (moral, social and sacrificial).

Those who choose Satan as eternal spiritual father and reject the gospel of eternal redemption back to God through faith and acceptance of Jesus Christ will spend eternity in a very unpleasant environment (first hell, then the lake of fire). All believers in Jesus

Christ become new living spiritual entities and strive to please God by behaving like Jesus.

All humans since Adam and Eve are living in eternity right now. Our human spirits breathed into Adam and Eve by God are immortal and therefore eternal. The entire human race including Adam and Eve and every individual born on Planet Earth, past, present and future, are either in heaven, hell, or still living on Earth in their physical bodies appointed to death. This summary is not the author's opinion nor private interpretation of God's plan for the redemption of humanity. It is absolute, unchangeable, and very specific Bible truth as set forth in the following Holy Scriptures:

"I am the God of Abraham, and the God

of Isaac, and the God of Jacob. God is not the
God of the dead, but of the living." (Matthew
chapter 22, verse 32 22 KJV)

"Then said Jesus again unto them, I go
My way, and ye shall seek Me, and shall die in
your sins: whither I go, ye cannot come."
(John, chapter 8, verse 21 KJV)

"I said therefore unto you, that ye shall
die in your sins: for if ye believe not that I am
He, ye shall die in your sins." (John, chapter
8, verse 24 KJV)

"Jesus said unto them, Verily, verily, I say
unto you, Before Abraham was, I AM." (John,
chapter 8, verse 58 KJV)

"Therefore doth My Father love Me,
because I lay down my life, that I might take it
again. No man taketh it from Me, but I lay it

down of Myself. I have power to lay it down, and I have power to take it again. This commandment have I received of My Father." (John, chapter 10, verses 17-18 KJV)

"I and My Father are one." (John, chapter 10, verse 30 KJV)

Chapter seven
Judgment Seat of Jesus Christ

The Judgment Seat of Jesus Christ and The Great White Throne Judgment are two separate divine judgments based upon entirely different criteria:

" For other foundation can no man lay than that is laid, which is Christ Jesus. Now if any man build upon this foundation gold, silver, precious stones, wood, hay, stubble; every man's work shall be made manifest: for the day shall declare it, because it shall be revealed by fire; and fire shall try every man's work of what sort it is. If any man's work abide which he hath built thereon, he shall receive a

reward. If any man's work shall be burned, he shall suffer loss: but he **himself** shall be saved; yet so as by fire." (I Corinthians, chapter 3, verses 11-14 KJV)

"But why does thou judge thy brother? or why does thou set at naught thy brother? For we shall all stand before the judgment seat of Christ. For it is written: As I live, saith the Lord, every knee shall bow to me, and every tongue shall confess to God. So then every one of us shall give account of himself to God." (Romans, chapter 14, verses 10-12 KJV)

"It is commonly reported that there is fornication among you, and such fornication as is not so much as named among the Gentiles, that one should have his father's wife. And ye are puffed up, and have not rather mourned,

that he that hath done this deed might be taken
away from among you. For I verily, as absent
in body, but present in spirit, have judged
already, as though I were present, concerning
him that hath so done this deed. In the name of
our Lord Jesus Christ, when ye are gathered
together, and my spirit, with the power of our
Lord Jesus Christ, to deliver such an one unto
Satan for the destruction of the flesh, that the
spirit may be saved in the day of the Lord
Jesus." (I Corinthians, chapter 5, verses 1-5
KJV)

"Fear none of those things which thou
shalt suffer: behold, the devil shall cast some
of you into prison, that ye may be tried; and ye
shall have tribulation ten days: be thou faithful
unto death, and I will give thee a crown of life.

He that hath an ear, let him hear what the spirit
saith unto the churches; He that overcometh
shall not be hurt of the **second death."**
(Revelation, chapter 2, verses 10-11 KJV)

"And he that overcometh, and keepeth
My works unto the end, to him will I give
power over the nations: and he shall rule them
with a rod of iron; as the vessels of a potter
shall they be broken to shivers: even as I
received of My Father. And I will give him the
morning star." (Revelation, chapter 2, verses
26-28 KJV)

"Behold, I come quickly: hold that fast
which thou hast, that no man take thy crown."
(Revelation, chapter 3, verses 10-11 KJV)

"And they sung a new song, saying, Thou
art worthy to take the book, and to open the

seals thereof: for Thou wast slain, and hast
redeemed us to God by Thy blood out of
every kindred, and tongue, and people, and
nation; and hast made us unto our God kings
and priests: and we shall reign on the earth."
(Revelation, chapter 5, verses 9-10 KJV)

"And the devil that deceived them was
cast into the lake of fire and brimstone, where
the beast and the false prophet are, and shall be
tormented day and night for ever and ever. And
I saw a **great white throne**, and Him that sat
on it, from Whose face the earth and the
heaven fled away; and there was found no
place for them. And I saw the dead, small and
great, stand before God; and the books were
opened: and another book was opened, which
is the book of life: and the dead were judged

out of those things which were written in the books, according to their works. And the sea gave up the dead which were in it; and death and hell delivered up the dead which were in them: and they were judged every man according to their works. And death and hell were cast into the lake of fire. This is the **second death.** And whosoever was not found written in the book of life was cast into the lake of fire." (Revelation, chapter 20, verses 10-15 KJV)

The "Judgment Seat of Christ" precedes the "Great White Throne Judgment" by one thousand years. The only humans who appear before the Judgment Seat of Christ are those who have believed in and accepted Jesus Christ as Lord and Savior and as their divine sacrificial lamb redeeming them back to the Godhead in a state of pure innocence. There is **no** guilt nor condemnation dispensed at the Judgment Seat of Christ; but rather the distribution by Jesus Christ of very precious rewards for faithful stewardship. Eternal life with Jesus is based on grace through faith and **not works.** Rewards are earned by faithful stewardship as an "Ambassador For Christ."

The only humans who appear at the Great White Throne Judgment are those who have exercised their God-given free will to rebel against God and choose Satan as their eternal spiritual father. They have in effect informed Satan and his demonic followers that God is their enemy and therefore they choose to follow Satan into his eternal habitat. At the

Great White Throne Judgment, Satan and all the fallen angels are judged **before humans** and cast forever into the eternal lake of fire and brimstone to be punished throughout eternity. Every immortal spiritual being that winds up in the lake of fire will be there because that is precisely what was chosen by an exercise of free will.

During the thousand years time window between the Judgment Seat of Christ and the Great White Throne Judgment, a series of events prophesied by John, the Apostle, while banished to the isle of Patmos by the Roman Empire in 96 A.D. will occur exactly as recorded by John in the Holy Scriptures:

1. The rapture of those "dead in Christ" and those "alive in Christ;"

2. The appearance of Satan "in the flesh" (the beast) following a war wherein Israel will exit as a world power;

3. The rise of an individual (the false prophet) who will organize all forms of worship into Satanism;

4. The beast elevated to global dictator in connection with a peace treaty to end all conflict in the Middle East and around the world;

5. The mark of the beast without which no one may buy or sell, and all who refuse the mark will be hunted down and tortured to death;

6.   The appearance of two "witnesses" who will defy the beast and shake the world with a series of devastating plagues, coupled with 144,000 individuals preaching the wrath of God upon the kingdom of the beast and salvation to those martyred during the reign of Satan;

7.   The battle of the nations (Armageddon) just prior to the return of Jesus Christ to Earth as "King of kings and Lord of lords;"

8.   Satan bound and powerless during the reign of Christ on

Earth (one thousand years);

9. Loosing of Satan and the final
    battle between God and Satan;

10. The judgment of all corrupted
    life forms on Earth and the "new
    creation" in conjunction with
    judgment of all who have
    rejected God's plan of
    redemption (Jesus); and
    eternal banishment of Satan's
    followers into the lake of fire.

The reign of Satan in the form of
Antichrist (the beast) will last seven years. The
reign of Christ on Earth will last one thousand
years; and time without end (eternity) will
commence thereafter. The trinity of Satan (the

dragon, the beast, and the false prophet) will spend eternity in the lake of fire joined by all of Satan's disciples (those who by choice selected Lucifer [Satan] as spiritual father, lord and master). Those humans who will occupy the lake of fire forever will be there by conscious choice. Because, when given the opportunity to choose between God and Satan, they chose the prince of darkness rather than the Sacrificial Lamb; they chose the lust of the flesh, the lust of the eyes and the pride of life rather than undefiled love, divine grace and tender mercy. They will be there because they deemed the free gift of redemption to be less desirable than the pleasures of sin for a season.

"........We have made a covenant with death, and with hell are we at agreement; when the overflowing scourge shall pass through, it shall not come unto us: for we have made lies our refuge, and under falsehood have we hid ourselves." (Isaiah, chapter 28, verse 15 KJV)

"The earth mourneth and fadeth away, the world languisheth and fadeth away, the haughty people of the earth do languish. The earth also is defiled under the inhabitants thereof; because they have transgressed the laws, changed the ordinance, broken the everlasting covenant. Therefore hath the curse devoured the earth, and they that dwell therein are desolate: therefore the inhabitants of the earth are burned, and few men left." (Isaiah, chapter 24, verses 4-6 KJV)

The following Holy Scripture has been referred to by believers in Jesus Christ as the

Rapture of the Church:

" But I would not have you to be ignorant, brethren, concerning them which are asleep, that you sorrow not, even as others who have no hope. For if we believe that Jesus died and rose again, even so them also which sleep in Jesus will God bring with Him. For this we say unto you by the word of the Lord, that we which are alive and remain unto the coming of the Lord shall not prevent them which are asleep. For the Lord Himself shall descend from heaven with a shout, with the voice of the archangel, and with the trump of God: and the dead in Christ shall rise first: then we which are alive and remain shall be caught up together with them in the clouds, to meet the Lord in the air: and so shall we ever be with the Lord."   (I Thessalonians, chapter 4, verses 13-18 KJV)

When the rapture occurs, the people who remain on Earth will be aware that certain Christians are missing, but will be unaware of the true reason therefor. Excuses will be offered to explain away the missing individuals who will represent a small percentage of Earth's human population. Life on Earth will continue and the **raptured** Christians will soon be old news as those left behind are captivated by emergence of a new celebrity (the Anti-Christ). This magnetic personality will provide believable solutions to previously insolvable problems and will orchestrate global peace including a treaty with Abraham's seed (Israel).

Pursuant to being accepted as dictator

over all humanity, the Antichrist will use the
absolute power given him to wage a genocide
campaign against all Jews and to hunt down
the "divine vomit" (left behind, lukewarm
Christians) who will recognize the Antichrist
as Satan in the flesh (the beast). The treaty
with Israel will be violated and the beast will
proclaim himself to be almighty god. Peace
will revert to global war ushering in pestilence,
famine, and decimation of Earth's population.

These events are revealed to John during
the portion of his Patmos vision pertaining to
"the things which shall be hereafter." A scroll,
written on both sides and sealed with seven
seals, is being unsealed by "the Lamb of God"
(Jesus). The angels are in attendance along
with four divinely ordained living creatures

(the four beasts), and twenty-four "elders" from among the raptured Christians and the redeemed **remnant** of Israel.

"..............the Lamb opened one of the seals, and I heard, as it were the noise of thunder, one of the four beasts saying, Come and see. And I saw, and behold a white horse: and he that sat on him had a bow; and a crown was given unto him: and he went forth conquering, and to conquer. And when He had opened the second seal, I heard the second beast say, Come and see. And there went out another horse that was red: and power was given to him that sat thereon to take peace from the earth, and that they should kill one another: and there was given unto him a great sword. And when He had opened the third seal, I heard the third beast say, Come and see. And I beheld, and lo a black horse; and he that sat on him had a pair of balances in his hand. And I heard a voice in the midst of the four beasts say, A measure of wheat for a penny, and three measures of barley for penny; and see thou hurt not the oil and the wine. And when He had opened the fourth seal, I heard the voice of the fourth beast say, Come and see. And I looked, and behold a pale horse: and his name that sat on him was Death, and Hell followed with him. And power was given unto them over the fourth part of the earth, to kill with sword, and with hunger, and with death, and with the beasts of the earth." (Revelation, chapters 4, a portion of verses 1-11; chapter 6, verses 1-8, Patmos vision, 96 A.D. KJV)

The nuclear war described in Zechariah, chapter 14, verse 12; and in Ezekiel, chapter 38, verses 1-23; chapter 39, verses 1-16 will set Israel apart from other nations to such an extent that the Antichrist will have to negotiate a peace treaty with the Jews in order to bring about the false peace which vaunts him into power as global dictator. This false peace will last forty-two months (the first half of the reign of Satan on earth). The peace will be broken when the beast (Antichrist) sets out to exterminate the seed of Abraham:

"And woe unto them that are with child, and to them that give suck in those days! But pray ye that your flight be not in winter, neither on the sabbath day: for then shall be great tribulation, such as was not since the beginning of the world to this time, no, nor ever shall be. And except those days should be shortened, there should no flesh be saved: but for the elect's sake those days shall be shortened." (Matthew, chapter 24, verses 19-22; KJV, Jesus Christ prophesying)

"And there followed him a great company of people, and of women, which bewailed and lamented him. But Jesus turning unto them said, Daughters of Jerusalem, weep not for Me, but weep for yourselves, and for your children. For, behold, the days are coming, in the which they shall say, Blessed are the barren, and the wombs that never bare, and the paps which never gave suck. Then shall they begin to say to the mountains, Fall on us; and to the hills, Cover us. For if they do these things in a green tree, what shall be done in the dry?" (Luke, chapter 23, verses 27-31; KJV, spoken by Jesus on the way to his sacrificial death)

In 538 B.C., the angel, Gabriel, explained the treachery of Satan in the flesh (Antichrist) to Daniel the prophet. The time references were given in "weeks of years" (one week equals seven years; or a total period of time equal to seventy times seven years):

"Seventy weeks are determined upon thy people and upon thy holy city, to finish the transgression, and to make an end of sins, and to make reconciliation for iniquity, and to bring in everlasting righteousness, and to seal up the vision and prophecy, and to anoint the

most Holy. Know therefore and understand,
that from the going forth of the commandment
to restore and to build Jerusalem unto Messiah
the Prince shall be seven weeks, and threescore
and two weeks: the street shall be built again,
and the wall, even in troublous times.   After
threescore and two weeks shall Messiah be cut
off, but not for Himself: and the people of the
prince that shall come shall destroy the city
and the sanctuary; and the end thereof shall be
with a flood, and unto the end of the war
desolations are determined. And he shall
confirm the covenant with many for one week:
and in the midst of the week he shall cause the
sacrifice and the oblation to cease, and for the
overspreading of abominations he shall make
it desolate......." (Daniel, chapter 9, verses 24-

27; spoken by Gabriel, 538 B.C. KJV)

Four hundred and eighty-three years (sixty-nine weeks of years) elapsed between the decree allowing the remnant of Judah and Benjamin to return to Jerusalem, and the death of Christ, exactly as predicted by Gabriel. One week of years (seven years) still remain to be fulfilled as referenced by Gabriel to Daniel in 538 B.C. This period of seven years will be fulfilled during the reign of the Antichrist, and the clock will begin ticking again **pursuant to the rapture of faithful Christians** from Earth. The first three and one half years of Anti-Christ's reign will be peaceful, and the final three and one half years will be filled with terror, torture, murder, war, and genocide:

"And one said to the man clothed in

linen, which was upon the waters of the river,

How long shall it be to the end of these

wonders? And I heard the man clothed in

linen, which was upon the waters of the river,

when he held up his right hand and his left

hand unto heaven, and sware by Him who

liveth forever that it shall be for a time, times,

and a half; and when he shall have accom-

plished to scatter the power of the holy people,

all these things shall be finished. And I heard,

but I understood not: then said I, O my Lord,

what shall be the end of these things? And he

said, Go thy way, Daniel: for the words are

closed up and sealed till the time of the end.

Many shall be purified, and made white, and

tried; but the wicked shall do wickedly; and

none of the wicked shall understand; but the

wise shall understand. And from the time that the daily sacrifice shall be taken away, and the abomination that maketh desolate set up, there shall be a thousand two hundred and ninety days."   (Daniel, chapter 12, verses 6-11 KJV)

The abomination that maketh desolate is the violation of the Jewish temple by Anti-Christ wherein Satan, in the flesh, occupies the temple and proclaims himself to be almighty god. This occurs at the midpoint of his seven year reign, and coincides with his efforts to kill every Jew, and to torture to death all who refuse to take his mark and to worship him as god. He will wage war against Israel, and roving death squads will hunt down the "divine vomit" (those lukewarm Christians **left behind at the rapture,** but knowing enough to reject Satan's mark and to refuse to worship him):

"And he had power to give life unto the image of the beast, that the image of the beast should both speak, and cause that as many as would not worship the image of the beast should be killed. And he causeth all, both small and great, rich and poor, free and bond, to receive a mark in their right hand, or in their foreheads: And that no man might buy or sell, save he had the mark, or the name of the beast, or the number of his name. Here is wisdom. Let him that hath under standing count the number of the beast: for it is the number of a man; and his number is six hundred threescore and six." (Revelation, chapter 13, verses 15-18; Patmos vision, 96 A.D. KJV)

To understand the rotating references to Satan, Antichrist, and the false prophet, it is helpful to remember that Satan is the spiritual power sustaining his incarnation in the body of the individual identified as the Antichrist, whereas the false prophet is the Satanic substitute for the Holy Spirit. The false prophet, acting as the world's spiritual figurehead, calls upon all humanity to worship the beast (Antichrist).

During the last forty-two months (3 1/2 years, 1,260 days) of Antichrist's reign, two witnesses oppose him and, like Moses opposing Pharaoh, call upon God to plague the kingdom of Antichrist and his followers. This period is also referred to as the "great tribulation." The population of Earth must endure both the wrath of God and the terror

perpetuated by Antichrist. The devastating

plagues pursuant to the ministry of the two

witnesses will be similar in nature to the

plagues suffered by Egypt during the time of

Moses, but of much greater intensity. The

mayhem, torture and murder wrought by

Antichrist will be of such magnitude that the

majority of Jews will be killed, and over half

of non-Jews will be slaughtered.

One hundred and forty-four thousand

Jewish male virgins will be protected by God

and serve as evangelists during the great

tribulation. Millions will be martyred rather

than worship Satan:

"After this I beheld, and lo, a great

multitude, which no man could number of all

nations, and kindreds, and people, and tongues,

stood before the throne, and before the Lamb,

clothed with white robes and palms in their

hands; And cried with a loud voice, saying,

Salvation to our God which sitteth upon the

throne, and unto the Lamb........And one of the

elders answered, saying unto me, What are

these which are arrayed in white robes? and

whence came they? And I said unto him, Sir,

thou knowest. And he said unto me, These are

they which came out of great tribulation, and

have washed their robes, and made them white

in the blood of the Lamb. Therefore are they

before the throne of God, and serve Him day

and night in His temple: and He that sitteth on

the throne shall dwell among them." (Reve-

lation, chapter 7, verses 9-10 and 13-15 KJV)

Toward the end of the second half of Antichrist's reign, an alliance of nations will rally themselves against Antichrist and his armies. The battle will commence in the Middle East at a place called Armageddon. Antichrist will be sitting in the Jewish temple as God Almighty. Armies opposing Antichrist will be attempting to dethrone him, and the armies supporting Antichrist will be defending his claim to divinity. The surviving Jews and the city of Jerusalem will be caught in the middle of the conflict. This final battle between humans orchestrated by Antichrist will be ended by the **second coming** of Jesus Christ:

"And I saw heaven opened, and behold a white horse; and He that sat upon him was

called Faithful and True, and in righteousness
He doth judge and make war. His eyes were as
a flame of fire, and on His head were many
crowns; and He had a name written, that no
man knew, but He Himself. And He was
clothed with a vesture dipped in blood: and
His name is called The Word of God. And the
armies which were in heaven followed Him
upon white horses, clothed in fine linen, white
and clean. And out of His mouth goeth a sharp
sword, that with it He should smite the
nations: and He shall rule them with a rod of
iron: and He treadeth the wine press of the
fierceness and wrath of Almighty God.......And
the beast was taken, and with him the false
prophet that wrought miracles before him, with
which he deceived them that had received the

mark of the beast, and them that worshiped his image. These both were cast alive into a lake of fire burning with brimstone. And the remnant were slain with the sword of Him that sat upon the horse, which sword proceeded out of His mouth (His spoken word): and all the fowls were filled with their flesh."
(Revelation, chapter 19, verses 11-15; 20-21 KJV)

The battle of Armageddon is followed by the reign of Christ on Earth which will encompass a period of one thousand years. The beast and the false prophet are imprisoned within the lake of fire, but Satan, himself, is bound and powerless during this period known as the "millennium reign of Christ." Earth, as we know it, will not be destroyed until the millennium reign is over, and the promises God made to King David, to Abraham, and to faithful Christians have been fulfilled by Jesus Christ sitting upon the throne of David for a millennium. Thereafter, Satan will be loosed and allowed to wage his final battle against God:

"And I saw an angel come down from heaven, having the key of the bottomless pit and a great chain in his hand. And he laid hold on the dragon, that old serpent, which is the Devil, and Satan, and bound him a thousand years, and cast him into the bottomless pit, and shut him up, and set a seal over him, that he deceive the nations no more, till the thousand years should be fulfilled: and after that he must be loosed a little season. And I saw thrones, and they sat upon them, and judgment was given unto them: and I saw the souls of them that were beheaded for the witness of Jesus, and for the word of God, and which had not worshiped the beast, neither his image, neither had received his mark upon their foreheads, or in their hands; and they lived and reigned with Christ a thousand years. But the **rest of the dead lived not again until the thousand years were finished.** This is the first resurrection. Blessed and holy is he that hath part in the first resurrection: on such the **second death hath no power,** but they shall be priests of God and of Christ, and shall reign with Him a thousand years. And when the thousand years are expired, Satan shall be loosed out of his prison, and shall go out to deceive the nations which are in the four quarters of the earth, Gog and Magog, to gather them together to battle: the number of whom is as the sand of the sea. And they went up on the breadth of the earth, and encompass- ed the camp of the saints about, and the beloved city: and fire came down from God out of heaven and devoured them. And the devil who deceived them was cast into the lake of fire and brimstone, where the beast and the false prophet **are,** and shall be tormented day and night for ever and ever." (Revelation, chapter 20, verses 1-10; Patmos vision, 96 A.D. KJV)

The consignment of Satan to the lake of fire is followed by the "great white throne" judgment. The only individuals who appear at this judgment are those who steadfastly refused to accept God's forgiveness through the divine sacrifice of Jesus Christ:

"And I saw a great white throne, and Him that sat on it, from Whose face the earth and the heaven fled away; and there was found no place for them. And I saw the dead, small and great stand before God; and the books were opened: and another book was opened, which is the book of life: and the dead were judged out of those things which were written in the books, according to their works. And the sea gave up the dead which were in it; and death and hell delivered up the dead which were in them: and they were judged every man according to their works. And death and hell were cast into the lake of fire. This is the second death. And whosoever was not found written in the book of life was cast into the lake of fire." (Revelation, chapter 20, verses 11-15 KJV)

Chapter 8
New Heaven and Earth

"But the day of the Lord shall come as a thief in the night; in the which the heavens shall pass away with a great noise, and the elements shall melt with fervent heat, the earth also and the works that are therein shall be burned up. Seeing then that all these things shall be dissolved, what manner of persons ought ye to be in all holy conversation and Godliness, looking for and hasting unto the coming of the day of God, wherein the heavens being on fire shall be dissolved, and the elements shall melt with fervent heat? Nevertheless we, according to His promise,

look for a new heavens and a new earth
wherein dwelleth righteousness."(II Peter,
chapter 3, verses 10-13 KJV)

"And I saw a new heaven and a new
earth: for the first heaven and the first earth
were passed away; and there was no more sea.
And I, John, saw the holy city, new Jerusalem,
coming down from God out of heaven,
prepared as a bride adorned for her husband.
And I heard a great voice out of heaven saying,
Behold the tabernacle of God is with men, and
He shall dwell with them, and they shall be His
people, and God Himself shall be with them,
and be their God. And God shall wipe away all
tears from their eyes; and there shall be no
more death, neither sorrow, nor crying, neither
shall there be any more pain: for the former

things are passed away. And He that sat upon
the throne said, Behold, I make all things new.
And He said unto me, Write: for these words
are true and faithful. And He said unto me, It is
done. I am Alpha and Omega, the beginning
and the end. I will give unto him that is athirst
of the fountain of the water of life freely. He
that overcometh shall inherit all things; and I
will be his God, and he shall be My son. But
the fearful, and unbelieving, and the abomin-
able, and murderers, and whoremongers, and
sorcerers, and idolaters, and all liars, shall have
their part in the lake of fire and brimstone:
which is the second death." (Revelation,
chapter 21, verses 1-8 KJV)

"And there came unto me one of the
seven angels which had the seven vials full of

the seven last plagues, and talked with me

saying, Come hither, I will shew thee the bride,

the Lamb's wife. And he carried me away in

the spirit to a great and high mountain, and

shewed me that great city, the holy Jerusalem,

descending out of heaven from God, having

the glory of God: and her light was like unto a

stone most precious, even like a jasper stone,

clear as crystal; and had a wall great and high,

and had twelve gates, and at the gates twelve

angels, and names written thereon, which are

the names of the twelve tribes of the children

of Israel: On the east three gates; on the north

three gates; on the south three gates; and on

the west three gates. And the wall of the city

had twelve foundations, and in them the names

of the twelve apostles of the Lamb. And he

that talked with me had a golden reed to measure the city, and the gates thereof and the wall thereof. And the city lieth foursquare, and the length is as large as the breadth: and he measured the city with the reed, twelve thousand furlongs. The length and the breadth and the height are equal. And he measured the wall thereof, an hundred and forty and four cubits, according to the measure of a man, that is, of the angel. And the building of the wall of it was of jasper: and the city was pure gold, like unto clear glass. And the foundations of the wall of the city were garnished with all manner of precious stones. The first foundation was jasper; the second sapphire; the third, a chalcedony; the fourth, an emerald; the fifth, sardonyx; the sixth, sardius; the seventh, chrysolyte; the eighth, beryl; the ninth, a topaz; the tenth, a chrysoprasus; the eleventh, a jacinth; the twelfth, an amethyst. And the twelve gates were twelve pearls; every several gate was of one pearl: and the street of the city was pure gold, as it were transparent glass. And I saw no temple therein: for the Lord God Almighty and the Lamb are the temple of it. And the city had no need of the sun, neither of the moon, to shine in it: for the glory of God did lighten it, and the Lamb is the light thereof. And the nations of them which are saved shall walk in the light if it: and the kings of the earth do bring their glory and honor into it. And the gates of it shall not be shut at all by day: for there shall be no night there. And they shall bring the glory and honor of the nations into it.

And there shall in no wise enter into it any thing that defileth, neither whatsoever worketh abomination, or maketh a lie: but they which are written in the Lamb's book of life. And he shewed me a pure river of water of life, clear as crystal, proceeding out of the throne of God and of the Lamb. In the midst of the street of it, and on either side of the river, was there the tree of life, which bare twelve manner of fruits, and yielded her fruit every month: and the leaves of the tree were for the healing of the nations." (Revelation, chapter 21, verses 1-27 and chapter 22, verses 1-2 KJV)

It is noteworthy that twelve thousand furlongs (12,000 times 582 feet per furlong divided by 5,280) equals 1,322.7 miles. One hundred and forty-four cubits (eighteen inches

per cubit) equals two hundred and sixteen feet.
The new Jerusalem will rise 1,322.7 miles
from the new Earth into the new heaven and
lies foursquare (1,322.7 x 1,322.7 x 1,322.7)
which will encompass two billion, three
hundred and fourteen million, one hundred
and ten thousand, three hundred and twenty-
eight cubic miles containing twelve trillion,
two hundred and eighteen billion, five hundred
and two million, five hundred and thirty
thousand cubit feet such that each one of the
twelve stories contains one trillion, eighteen
billion, two hundred and eight million, five
hundred and forty-four thousand cubit feet.

The scriptures pertaining to the new
heaven and the new Earth state that the nations
and the kings of the Earth will bow to the

glory and honor of new Jerusalem which will serves as the seat of God's sovereignty as God Himself dwells among His human sons and daughters.

The thousand reign of Jesus Christ on Planet Earth prior to the new heaven and the new Earth will involve nations, peoples and tongues that survived the Battle of Armageddon and the global genocide efforts of the Antichrist during his seven year reign over humanity. Death will continue to be the foremost enemy of humans but the lifespan of those who enter the reign of Christ in their natural bodies will be greatly extended. One who lives an hundred years will be considered a mere child. The believers in Jesus as their Sacrificial Lamb, Lord and Savior who were raptured from Earth prior to the seven year reign of Satan on Earth will rule and reign alongside Jesus Christ:

"And I will pour upon the house of David, and upon the inhabitants of Jerusalem, the spirit of grace and of supplications: and they shall look upon Me Whom they have pierced, and they shall mourn for Him, as one mourneth for his only son, and shall be in bitterness for Him, as one that is in bitterness for his firstborn." (Zechariah, chapter 12, verse 10   KJV)

"And it shall come to pass, that every one that is left of all the nations which came against Jerusalem shall even go up from year to year to worship the King, the Lord of Hosts, and to keep the feast of tabernacles. And it shall be, that whoso will not come up of all the families unto Jerusalem to worship the King, the Lord of Hosts, even upon them shall be no rain." (Zechariah, chapter 14, verses 16-17 KJV)

"For, behold, I create new heavens and a new Earth: and the former shall not be remembered, nor come into mind. But be ye glad and rejoice for ever in that which I create: for, behold, I create Jerusalem a rejoicing, and her people a joy. And I will rejoice in Jerusalem, and joy in My people: and the voice of weeping shall be no more heard in her, nor the voice of crying. There shall be no more thence an infant of days, nor an old man that hath not filled his days: for the child shall die an hundred years old; but the sinner being an hundred years old shall be accursed." (Isaiah, chapter 65, verses 17-20 KJV)

"For as the new heavens and the new Earth, which I will make, shall remain before Me, saith the Lord, so shall your seed and your name remain. And it shall come to pass, that from one new moon to another, and from one sabbath to another, shall all flesh come to worship before Me, saith the Lord." (Isaiah, chapter 66, verses 22-23 KJV)

"And I saw thrones, and they sat upon them, and judgment was given unto them: and I saw the souls of them that were beheaded for the witness of Jesus, and for the word of God, and which had not worshiped the beast, neither his image, neither had received his mark upon their foreheads, or in their hands; and they lived and reigned with Christ a thousand years. But the rest of the dead lived not again until the thousand years were finished. This is the first resurrection. Blessed and holy is he that hath part in the first resurrection: on such the second death hath no power, but they shall be priests of God and of Christ, and shall reign with Him a thousand years." (Revelation, chapter 20, verses 4-6 KJV)

Bibliography

Other than the Authorized King James Version of the Holy Bible, the author has not directly quoted from the titles listed in the bibliography. The referenced titles adequately cover the belief in special creation versus the hypotheses of Darwinian evolution.

Carbon-14 Dating, Radiometric Dating and Tree Ring Dating

[1] Plastino, W.; Kaih ola, L.; Bartolomei, P.; Bella, F. (2001). "Cosmic Background Reduction In The Radiocarbon Measurement By Scintillation Spectrometry At The

Underground Laboratory Of Gran
Sasso".

[2] Arnold, J. R.; Libby, W. F. (1949).
"Age Determinations by Radiocarbon
Content: Checks with Samples of
Known Age". _Science_ **110** (2869): 678–
680. doi:10.1126/science.110.2869.678.
PMID 15407879.

[3] Willard Frank Libby
Münnich KO, Östlund HG, de Vries H
(1958). "Carbon-14 Activity during the
past 5,000 Years". _Nature_ **182** (4647):
1432–3. doi:10.1038/1821432a0.

[4] Ramsey, C. Bronk (2008).
"Radiocarbon dating: revolutions in

understanding". _Archaeometry_ **50** (2): 249-275. doi:10.1111.2Fj.1475-4754.2008.00394.x. edit

[5] Scott, EM (2003). "The Fourth International Radiocarbon Intercomparison (FIRI).". _Radiocarbon_ **45**: 135–285.

[6] "NOSAMS Radiocarbon Data and Calculations". Woods Hole Oceanographic Institution. http://www.nosams.whoi.edu/clients/data.html.

[7] Stuiver M, Reimer PJ, Braziunas TF (1998). "High-precision radiocarbon age calibration for terrestrial and marine

samples". *Radiocarbon* 40: 1127–51.
http://depts.washington.edu/qil/datasets/
uwten98_14c.txt.

[8]   Suter M, Wölfli W (1994).
"Systematic investigation of
uncertainties in radiocarbon dating due
to fluctuations in the calibration curve".
*Nuclear Instruments and Methods in
Physics Research* **92**: 194–200.
doi:10.1016/0168-583X(94)96004-6.

[9]   Lerman, J. C.; Mook, W. G.; Vogel,
J. C.; de Waard, H. (1969). "Carbon-14
in Patagonian Tree Rings". *Science* **165**
(3898): 1123–1125.

doi:10.1126/science.165.3898.1123.
PMID 17779805.

[10] Kolchin BA, Shez YA (1972). *Absolute archaeological datings and their problems*. Moscow: Nauka.

[11] Beck JW; Richards, DA; Edwards, RL; Silverman, BW; Smart, PL; Donahue, DJ; Hererra-Osterheld, S; Burr, GS et al. (2001). "Extremely large variations of atmospheric C-14 concentration during the last glacial period.". *Science* **292** (5526): 2453–2458. doi:10.1126/science.1056649. PMID 11349137.

[12] Hoffmann DL; Beck, J. Warren; Richards, David A.; Smart, Peter L.;

Singarayer, Joy S.; Ketchmark, Tricia; Hawkesworth, Chris J. (2010). "Towards radiocarbon calibration beyond 28 ka using speleothems from the Bahamas". *Earth and Planetary Science Letters* **289**: 1–10. Bibcode 2010E&PSL.289....1H. doi:10.1016/j.epsl.2009.10.004.

[13] Pennicott K (10 May 2001). "Carbon clock could show the wrong time". *PhysicsWeb*. http://physicsworld.com/cws/article/news/2676.

Big Bang Theory

[14] D. N. Spergel et al. (2007). "Three-Year Wilkinson Microwave Anisotropy Probe (WMAP) Observations: Implications for Cosmology". *Astrophysical Journal Supplement Series* **170** (2): 377–408. arXiv:astro-ph/0603449. Bibcode 2007ApJS..170..377S. Doi:10.1086/513700.

[15] Liddle, Andrew; David Lyth (2000). *Cosmological Inflation and Large-Scale Structure*. Cambridge. ISBN 0-521-57598-2.

[16] Edmund Bertschinger (1998).

"Simulations of structure formation in the universe". *Annual Review of Astronomy and Astrophysics* **36** (1): 599–654. Bibcode 1998ARA&A..36..599B. doi:10.1146/annurev.astro.36.1.599.

[17] Harrison, E. R. (1970). "Fluctuations at the threshold of classical cosmology". *Phys. Rev.* **D1**: 2726. Bibcode 1970PhRvD...1.2726H. doi:10.1103/PhysRevD.1.2726.

[18] Peebles, P. J. E.; Yu, J. T. (1970). "Primeval adiabatic perturbation in an expanding universe". *Astrophysical Journal* **162**: 815. Bibcode

1970ApJ...162..815P.

Doi:10.1086/150713.

[19] Ya; Zel'dovich, B. (1972). "A hypothesis, unifying the structure and entropy of the universe". *Monthly Notices of the Royal Astronomical Society* 160. Bibcode 1972MNRAS.160P...1Z.

[20] R. A. Sunyaev, "Fluctuations of the microwave background radiation," in *Large Scale Structure of the Universe* ed. M. S. Longair and J. Einasto, 393. Dordrecht: Reidel 1978.

Quantum Mechanics

[21]   "On the Law of Distribution of Energy in the Normal Spectrum". Francis Weston Sears (1958). *Mechanics, Wave Motion, and Heat*. Addison-Wesley. p. 537. http://books.google.com/books?hl=en&q=%22Mechanics%2C+Wave+Motion%2C+and+Heat%22+%22where+n+%3D+1%2C%22&btnG=Search+Books.

[22] Kragh, Helge (1 December 2000). "Max Planck: the reluctant revolutionary". PhysicsWorld.com. http://physicsworld.com/cws/article/prin

[22] McEvoy, J. P.; Zarate, O. (2004).
Introducing Quantum Theory. Totem
Books. pp. 70–89.
[23] Dicke and Wittke, *Introduction to
Quantum Mechanics*, p. 10f.

Theory of Relativity

[24] Einstein A. (1916 (translation
1920)), Relativity: The Special and
General Theory, New York: H. Holt and
Company

[25] Miller, Arthur I. (1981), *Albert
Einstein's special theory of relativity.*

Emergence (1905) and early interpretation (1905–1911), Reading: Addison–Wesley, <u>ISBN</u> <u>0-201-04679-2</u>

[26] Will, Clifford M (August 1, 2010). <u>"Space-Time Continuum"</u>. *Grolier Multimedia Encyclopedia.*

[27] Einstein's letter to the London Times in 1919.

- Einstein Albert (Nov. 28, 1919). <u>"What is the theory of relativity?"</u>

The foregoing bibliography is not actually associated with direct quotes from the reference material, but rather was researched to comprehend the 21st century scientific disciplines indicated as opposed to the 19th century scientific literature on the same subject matter.